For John Bradley King –
one of my favourites –
from his aunt –
Marian King –
with much love –
Oct. 1978

ADVENTURES IN ART

NATIONAL GALLERY OF ART, WASHINGTON, D.C.

ADVENTURES

NATIONAL GALLERY OF ART
WASHINGTON, D.C.

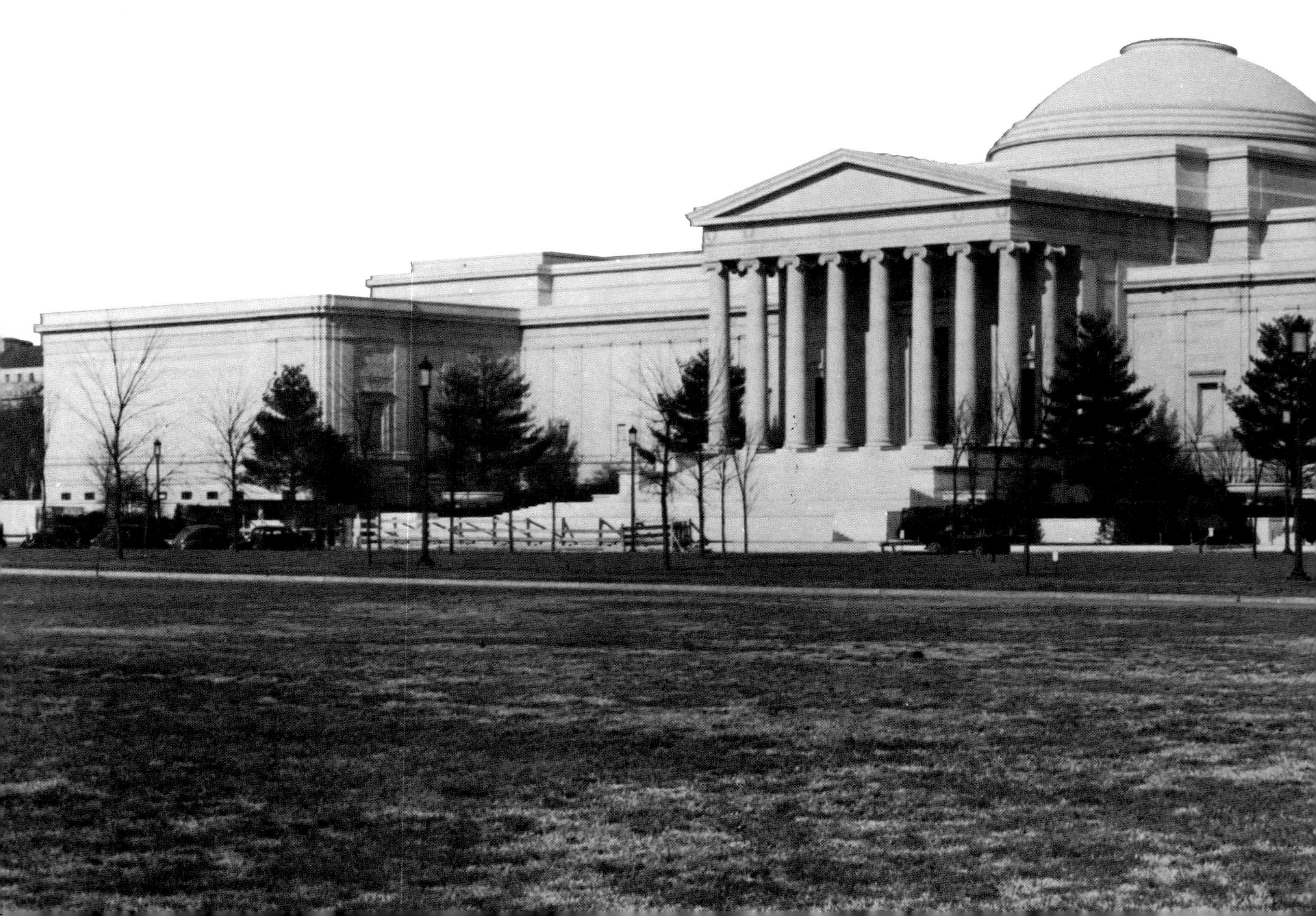

Marian King

IN ART

HARRY N. ABRAMS, INC., PUBLISHERS, NEW YORK

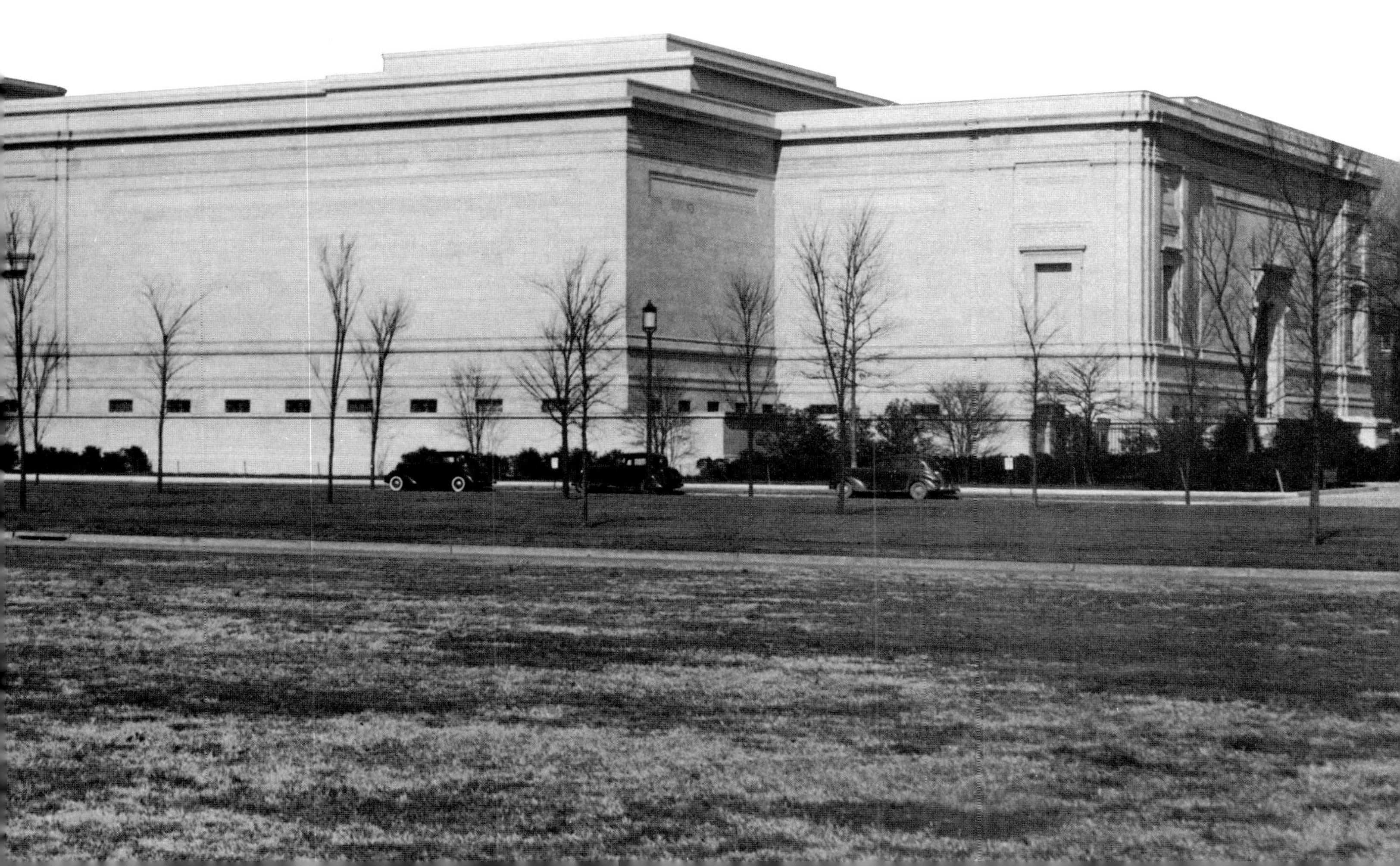

Note: The commentaries on pages 54, 64, 73, 88 and 105 are adapted from material previously published in *Highlights for Children*, copyright © 1972, 1974, 1975, 1976 Highlights for Children, Inc., Columbus, Ohio

Editor: Patricia Gilchrest
Assistant Editor: Ellen Schwartz
Designer: Deborah Jay

Library of Congress Cataloging in Publication Data
Main entry under title:

Adventures in art.

Bibliography: p.
Includes index.
1. Art—Washington, D.C. 2. United States. National Gallery of Art. I. King, Marian. II. United States. National Gallery of Art.
N856.A513 750'. 74'0153 77-20926
ISBN 0-8109-1769-6
ISBN 0-8109-2167-7

Library of Congress Catalogue Card Number: 77-20926

Published in 1978 by Harry N. Abrams, Incorporated, New York

Printed and bound in Japan

To Paul Mellon—with warm appreciation
and
In fond remembrance of my good friends
David E. Finley
and
William Patrick Campbell

CONTENTS

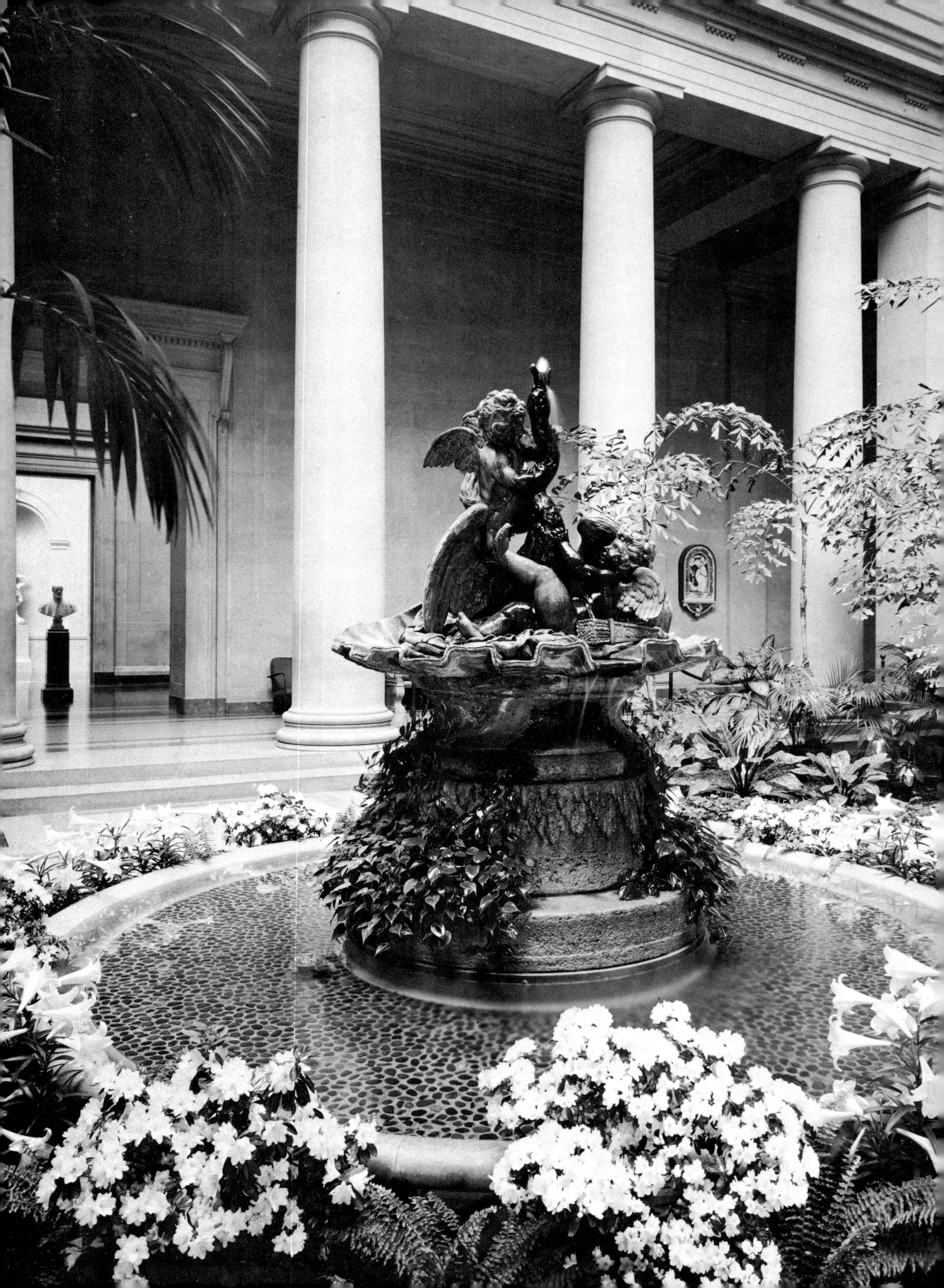

FOREWORD

The National Gallery of Art, the youngest of the world's great galleries, presents in its collections many of the greatest examples of Western European and American works from the twelfth century to the present—all the result of gifts from private individuals, all intended for the pleasure and enlightenment of people throughout the world. Given such a time-span, the number of great works of art available, and the variety of nationalities, schools of painting, individual artists, and particular paintings from which to choose, a person confronting this variety and richness cannot help but feel a little awed by it all.

This is where a book such as Marian King's can open up a new world of discovery. Her selection ranges across the full gamut of the Gallery's holdings. Each picture is treated with lively appreciation and a panoply of facts.

The pages that follow form a seminal and informative guide. The use of it before a visit, after a visit, during a visit, or simply by itself cannot help but bring enrichment and the joy of discovery.

J. Carter Brown
Director
National Gallery of Art

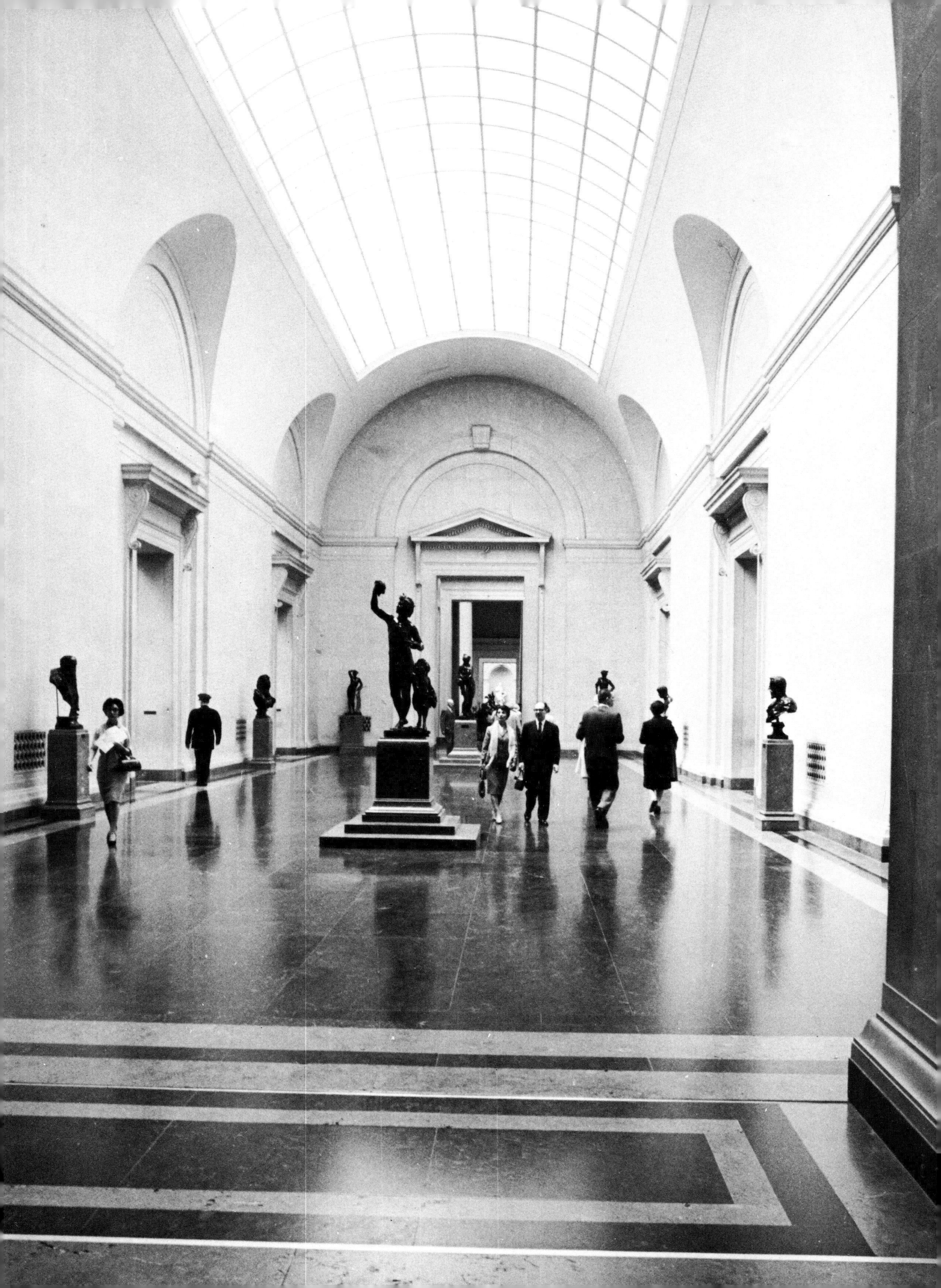

INTRODUCTION

The National Gallery of Art in Washington, D.C., belongs to all of us, just as do the Capitol building and the Lincoln Memorial. This great museum was the gift to the nation of Andrew W. Mellon, a public-spirited man who loved great works of art.

When Mr. Mellon was this country's Ambassador to Great Britain, he often visited London's National Gallery, which he greatly admired. He began to make plans for such a repository of art in the United States. Although he had been collecting paintings since he was a young man, Mr. Mellon now began to buy in a more concentrated fashion. His aim was to acquire the finest art that could be found for the new museum. Once the gallery was established, he hoped that it would be the recipient of fine works of art donated by other collectors.

In a letter to President Franklin D. Roosevelt, written three days before Christmas in 1936, Mr. Mellon formally promised to provide funds to build the gallery and offered to donate the collection of works of art he had acquired. He specified that he did not want the museum to be named after him; rather it was to have a name that would identify it as belonging to the people of the United States. It was to be called, therefore, the National Gallery of Art.

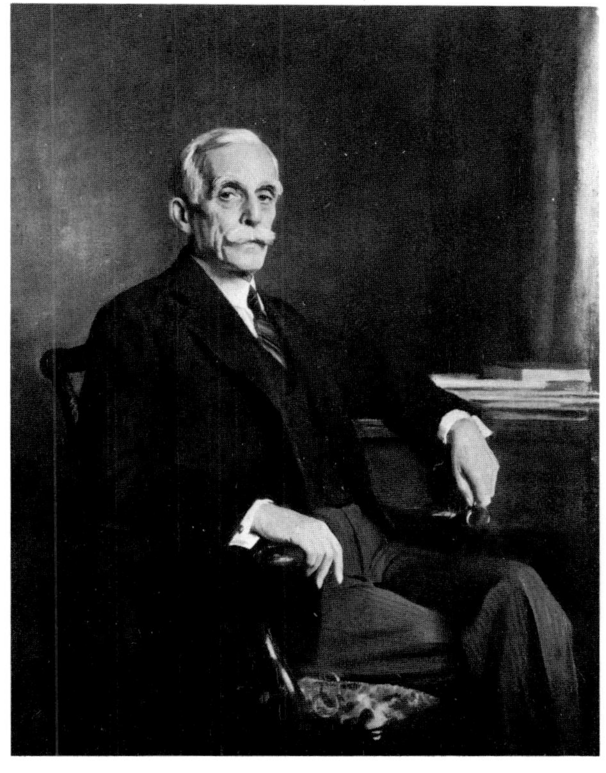

Andrew Mellon. Portrait by Oswald Birley. Gift of Mrs. Mellon Bruce.

In answering Mr. Mellon's letter, President Roosevelt said, "I was not only completely taken by surprise but was delighted by your wonderful offer. . . . This was especially so because for many years I have felt the need for a national gallery of art in the Capital."

On March 24, 1937, President Roosevelt approved the statute which established the Gallery. In June of that year ground was broken and work started on the building. The distinguished architect John Russell Pope designed the building in the Neo-Classical style. This style was chosen in order to reflect the timelessness of the art as well as to integrate the building with the surrounding Neo-Classical architecture. It took nearly four years to complete this monumental domed structure with its fourteen different types of marble and a variety of other decorative stones. The exterior of the Gallery is of rose-white Tennessee marble.

Unfortunately, Andrew Mellon did not live to see his glorious dream fulfilled. He died in 1937, soon after work on the building was begun. On March 17, 1941, the heavy bronze doors of the National Gallery of Art were opened to the people. President

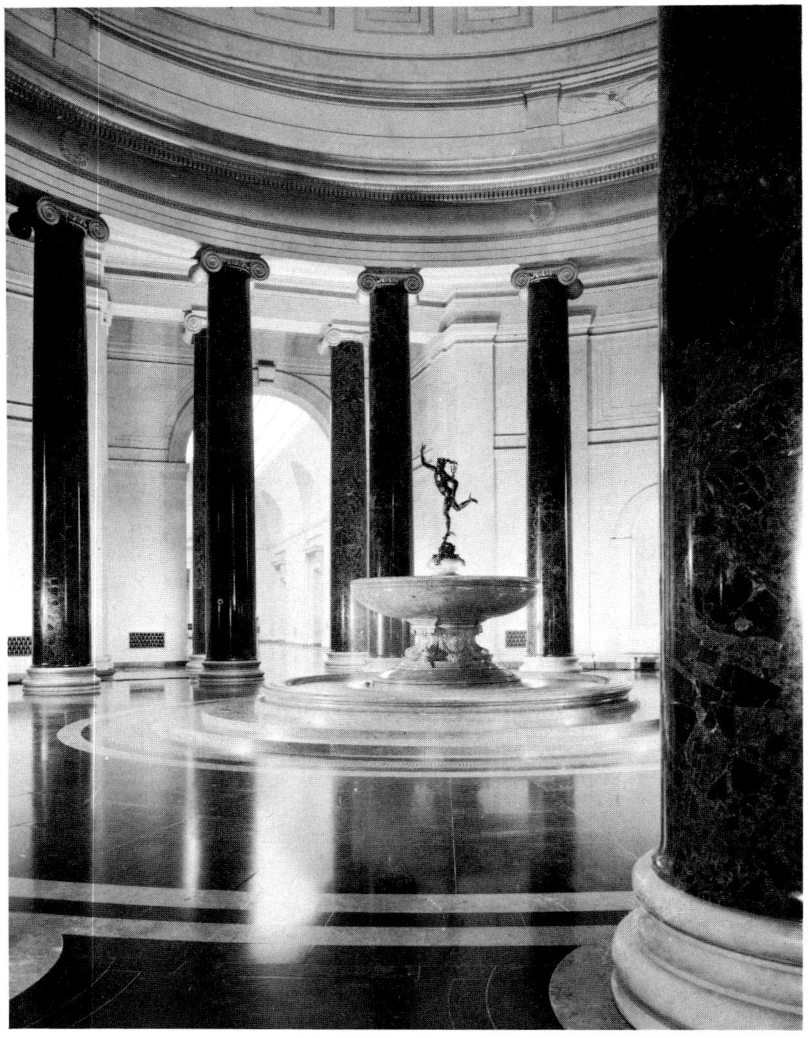

View of the Rotunda of the National Gallery of Art, showing the bronze statue of Mercury.

Roosevelt accepted Mr. Mellon's gracious gift on behalf of the nation.

The visitor to the National Gallery of Art who uses the Washington Mall entrance will see as he reaches the main floor the simple but grand plan of the building: a rotunda with two wings. The first view upon entering is the breathtaking circle of massive dark green Italian marble columns, over thirty-six feet high, which help define the shape of the rotunda. In the center of the ring of columns the graceful bronze figure of Mercury stands high above a fountain with gently splashing water. Radiating out from the fountain is a floor composed of green marble from Vermont and gray marble from Tennessee.

The decorative sunken panels, or coffers, in the dome of the rotunda and the opening in its top are an adaptation of the interior of the Pantheon, one of the most famous architectural monuments of ancient Rome. This circular temple of the second century

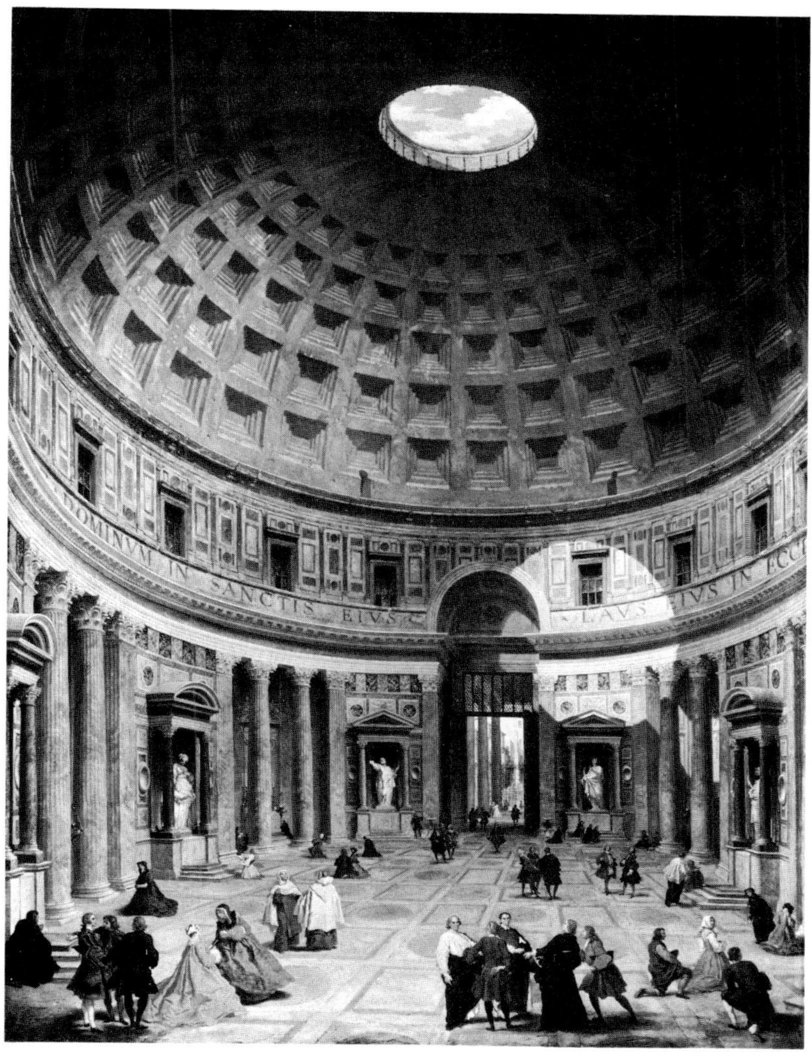

The Pantheon, the inspiration for the central rotunda of the National Gallery, as depicted in Giovanni Paolo Panini's painting of c. 1740, owned by the Gallery. Samuel H. Kress Collection.

B.C., however, was open to the sky; in the National Gallery a glass skylight covers the opening. In the Pantheon the distance from the floor to the top of the dome is the same as the diameter of the interior: 143 feet. In the gallery these ideal proportions have been slightly altered: the 103-foot ceiling is two feet higher than the interior's 101-foot diameter.

Stretching out to the east and west from the rotunda are long halls lined with sculpture. Opening off these halls are the many rooms which house the Gallery's collection. At the end of each hall is a garden court leading to other gallery rooms. In the center of each court stands a seventeenth-century fountain from the Théâtre de l'Eau at Versailles.

On Sunday evenings, the National Gallery Orchestra or guest musicians give free concerts in the East Garden Court.

Andrew Mellon's collection, which formed the basis of the National Gallery of Art, consisted of 126 paintings and 26 pieces of

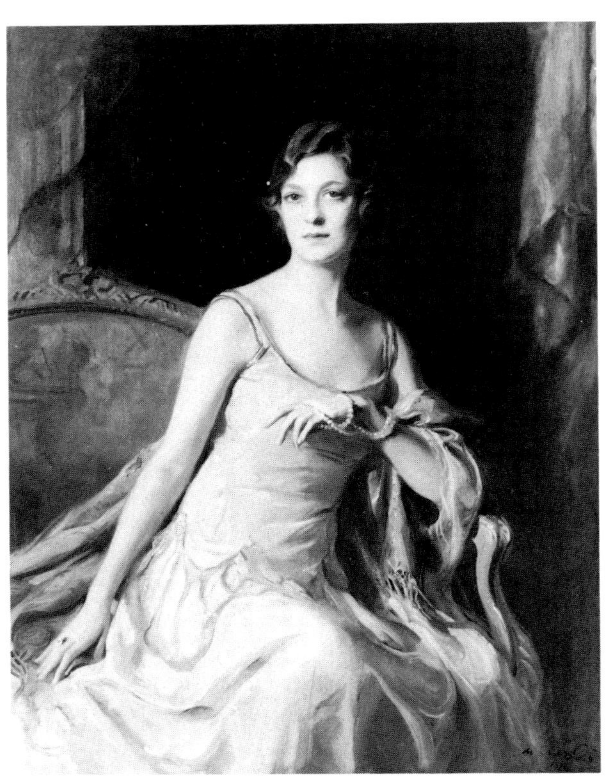

Ailsa Mellon Bruce. Portrait by Philip de Lászlo. Collection of Ailsa Mellon Bruce.

sculpture. These works of art included such masterpieces as Raphael's *Alba Madonna, Niccolini-Cowper Madonna,* and *St. George and the Dragon*; Van Eyck's *Annunciation*; Botticelli's *Adoration of the Magi*; and nine paintings by Rembrandt. Mr. Mellon's hope that other collectors would give their masterpieces to the Gallery has been magnificently realized, and today the Gallery's collection continues to be augmented by private donations. More than two hundred donors have contributed to its growth. Federal funds, although used for the museum's maintenance and operating costs, are not available for the purchase of works of art; the growth of the Gallery's collections is therefore dependent on the generosity of its benefactors.

Even before the National Gallery opened, Samuel H. Kress donated his great collection of Italian paintings and sculpture dating from the thirteenth to the eighteenth century. Later, he and his brother, Rush H. Kress, enlarged the collection with more works of Italian painting and sculpture as well as early Flemish and German paintings and important works from other schools of painting. Among the paintings in the collection are *The Adoration of the Magi,* the great tondo (a circular painting) by Fra Angelico and Fra Filippo Lippi, and El Greco's *Laocoön.*

In 1942 Joseph E.Widener donated the collection of paintings, sculpture, and decorative arts—tapestries, Renaissance jewelry, Chinese porcelains, and fine examples of eighteenth-century French furniture—which he and his father, P.A.B. Widener, had formed. The Widener Collection is highlighted by fourteen Rembrandts, eight Van Dycks, and two Vermeers.

Chester Dale helped to fill the Gallery in the 1940s and 1950s by lending his masterpieces of American and French painting, which included the beloved *Girl with a Watering Can* by Auguste Renoir. At Mr. Dale's death in 1962, these loans became gifts. He also left a large bequest for purchases and fellowships.

Adding to the Gallery's collection has often been a family affair. Andrew Mellon's daughter, the late Ailsa Mellon Bruce, contributed funds in order to round out the collection through purchases of important works of art as they came on the market. Among the great works of art bought were *Ginevra de'Benci,* the only generally acknowledged painting by Leonardo da Vinci in America; Georges de la Tour's *Repentant Magdalen;* and Picasso's *Nude Woman.* Andrew Mellon's son, Paul Mellon, the President of the National Gallery since 1963, has given as well as lent many French, British, and American paintings.

The National Gallery's Graphic Arts Department owes its beginnings to Lessing J. Rosenwald's gift of over twenty thousand prints and drawings from his superb private collection. Added to by many other donors and by purchases, the graphic arts collection now contains about fifty thousand examples ranging from the fifteenth century to the present day. The prints, drawings, manuscript illuminations, and book illustrations are displayed on a rotating basis.

Another very special collection is the Index of American Design, which contains over 17,000 watercolor renderings of American crafts and folk arts from before 1700 to about 1900.

In addition to works in the Gallery's own collections, works of art from private collections and from museums all over the world are frequently displayed in special exhibitions.

The Education Department of the Gallery has a large staff of lecturers and docents who give talks and conduct tours on a regular basis. The department is also responsible for interpreting to the public the Gallery's special exhibitions through wall labels, recorded tours, films, and taped slide lectures. It has an art information service which maintains the information desks and answers questions from the public on all aspects of art history and the Gallery's collections, and a slide library for the use of scholars and the general public which contains over 75,000 slides, including works not in the Gallery.

Educational audio-visual materials—films and slide programs—are circulated free of charge through the Gallery's extension series to more than four thousand communities in all fifty states and several foreign countries. The borrowers include schools, government agencies and museums, armed services education centers, libraries, art organizations, and private individuals.

Thus far, the National Gallery has had three directors: David E. Finley, John Walker, and J. Carter Brown.

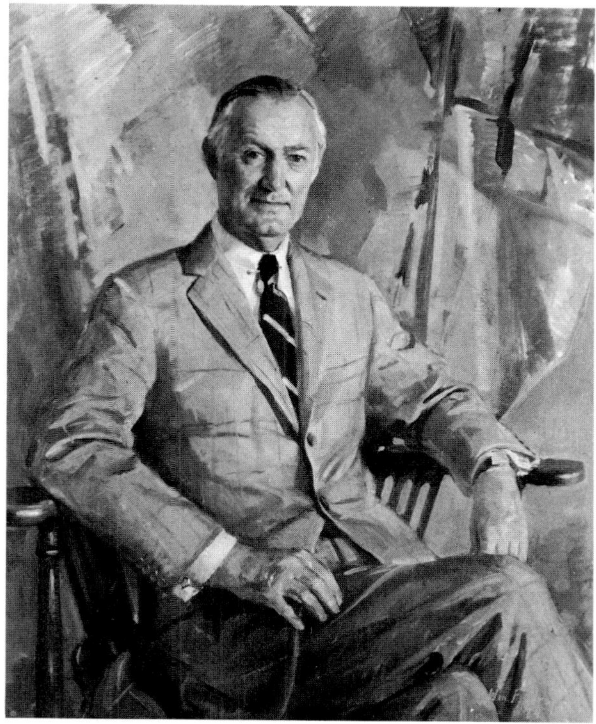

Paul Mellon. Portrait by William Draper.

A new addition to the National Gallery, the East Building, has been made possible through the generosity of Paul Mellon and the late Ailsa Mellon Bruce.

At the ground-breaking ceremony, on May 6, 1971, Paul Mellon, the President of the Gallery, said: "I am pleased to have you witness with us a ceremony that symbolizes perhaps the most important single forward step since the original Gallery building opened to the public in 1941. On a less official level, I am delighted to have you share in a moment that brings closer a goal for which many have worked and in which I have deep personal interest."

The form of the East Building was determined by the shape of its site, a trapezoidal piece of land between the Capitol and the original National Gallery building. The site is bordered by the diagonal of Pennsylvania Avenue on the north and the Mall on the south. The architect for the project, I. M. Pei, designed a structure in which unusual and dramatic triangular forms are used to repeat the angles formed by the trapezoid. The building looks like an enormous piece of modern sculpture, which is appropriate for a museum which displays many modern works of art among its exhibits. The spirit of the building is in harmony with its contents

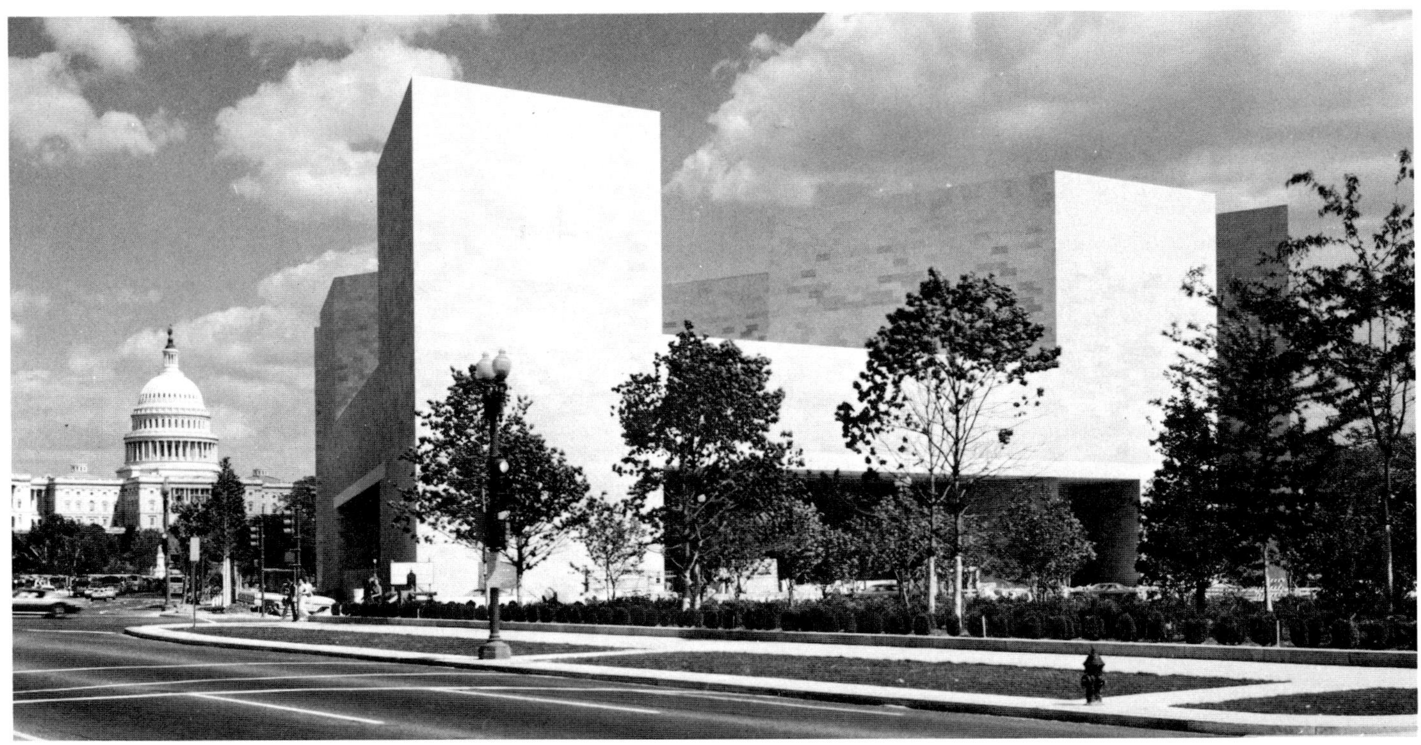

East Building of the National Gallery of Art.

just as the Neo-Classical architecture of the original building reflects the Classical influences which appear in much of the art it contains.

Among the many works of modern art which have found a home in the East Building are Henry Moore's *Knife-Edge, Mirror, Two Piece* on the entry terrace, a group of Matisse's large collages of cut paper, Picasso's *Family of Saltimbanques*, and Mondrian's *Lozenge in Red, Yellow and Blue*. Also on view are modern tapestries: *Variation sur Aubette* by Jean Arp and *Femme* by Joan Miró. A recently commissioned, mural-sized painting by the contemporary American artist Robert Motherwell hangs beneath the Terrace Gallery skylight; and a giant tricolored, multipart mobile, one of the late Alexander Calder's last works, hangs from the glass roof of the spacious interior courtyard.

The East Building houses a Center for Advanced Study in the Visual Arts, comprising among other tools for research an enlarged library with space for over 300,000 volumes, and an invaluable photographic archive which is constantly being supplemented and improved through grants from the Kress Foundation.

Linking the new building with the old is a four-acre plaza with an underground concourse area. The central structure in the plaza is a combination fountain-skylight which provides daylight and animation to the "sidewalk" café in the concourse. Such additions promise to continue the National Gallery's tradition of providing the setting which best enhances the visitor's appreciation of art's greatest achievements.

THE PLATES

1. ANDREA DEL CASTAGNO. *The Youthful David.* c. 1450. Leather, 45 1/2 × 30 1/4″. Widener Collection

ANDREA DEL CASTAGNO

The Youthful David

During the Renaissance in Italy, festive decorated shields, painted on leather stretched over wooden braces and used for ceremonial parades, were popular. Andrea del Castagno's *The Youthful David* is the only such shield by a great master still in existence. The painting is a symbolic representation of the Old Testament story of the Israelite shepherd boy David and the Philistine giant Goliath. To end a stalemated war, Goliath challenged any Israelite to meet him in single combat, saying, "Choose you a man . . . and let him come down to me. If he be able to fight with me, and to kill me, then will we be your servants: but if I prevail against him, and kill him, then shall ye be our servants and serve us. . . . I defy the armies of Israel this day; give me a man, that we may fight together."

It is David's triumph over Goliath with a slingshot as his only weapon that Castagno portrays. Silhouetted against a deep blue sky flecked with decorative clouds, David, graceful and agile, stands on jagged rocks in front of spiky green foliage. His purple-red tunic is blown in graceful folds by the wind. His face, framed by brown flowing hair, is turned in the direction of his target. His eyes are tense, his expression intent. He grasps the loaded sling in his right hand; his left hand is raised to sight the course of his missile. On the ground between his feet lies the bloodied head of Goliath, which David severed after knocking the giant down. Thus David's act and its outcome are shown at the same time.

The painting reveals the interest of the artists of the Renaissance in the anatomy of the human body. The figure of David is strongly modeled, like sculpture. Beneath the skin the artist suggests the underlying muscles of the legs, the bony structure of the head, the tendons of the neck, and the veins of the arms.

To the Florentine people of Castagno's time the victory depicted on the shield was especially meaningful: it represented their own political liberty.

Andrea del Castagno

Andrea del Castagno took his name, as did many Renaissance artists, from the town of his birth, Castagno, Italy. He was born about 1417–19, and spent most of his life in nearby Florence. His work was sought after by both public and private patrons, for whom he executed many religious pictures as well as portraits villa decorations, and ceremonial objects such as the shield depicting David and Goliath. Castagno died from the plague which swept Florence in 1457.

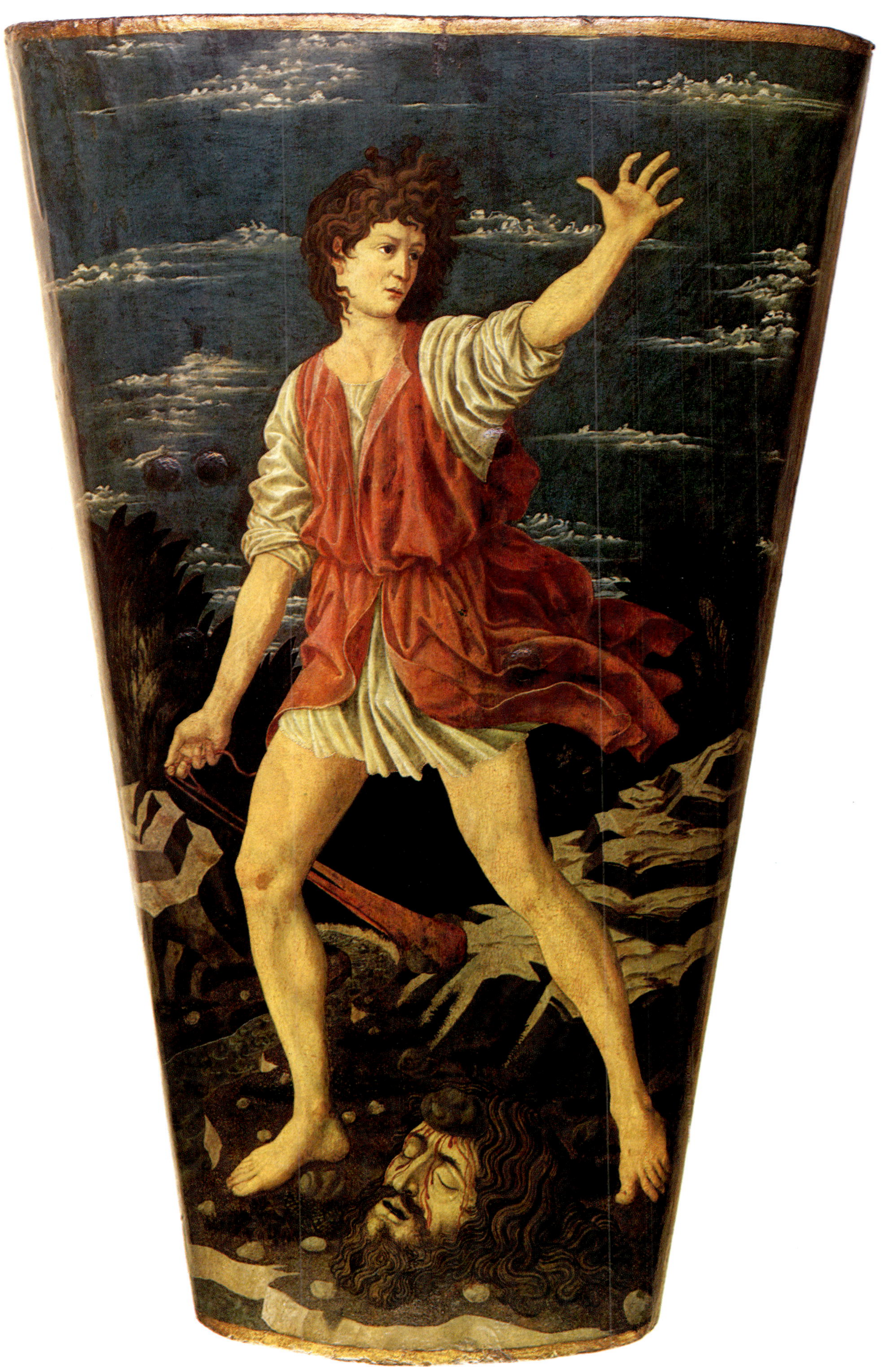

FRA ANGELICO
AND FRA FILIPPO LIPPI

The Adoration of the Magi

Now when Jesus was born in Bethlehem of Judea in the days of Herod the king, behold, there came wise men from the east to Jerusalem, Saying, Where is he that is born King of the Jews? for we have seen his star in the east, and are come to worship him.

Then Herod . . . enquired of them diligently what time the star appeared. And he sent them to Bethlehem, and said, Go and search diligently for the young child; and when ye have found him, bring me word again that I may come and worship him also. When they heard the king, they departed; and, lo, the star which they saw in the east, went before them, till it came and stood over where the young child was. When they saw the star, they rejoiced with exceeding great joy. And when they were come into the house, they saw the young child with Mary his mother, and fell down, and worshipped him: and when they had opened their treasures, they presented unto him gifts; gold, and frankincense, and myrrh.

The focal point of this circular painting, or tondo, by two monks is the seated Virgin, clad in a pink dress and blue mantle, holding the Christ Child.

Beside Mary, his hand raised in homage, stands Joseph, in a pale gold robe, blue tunic and headdress. His position and gesture are echoed by the rudely clothed figure behind him. Halos, symbols of divinity, adorn the heads of the Holy Family.

The oldest of the Magi, in lavender, kneels in adoration before the Virgin and Child. Behind him are the two other wise men. One wears an embroidered scarlet garment and offers a golden vessel of myrrh; the other, in blue and pale yellow, kneels on a carpet of flowers, the decorative pattern of which recalls medieval tapestries. In the foreground of the painting is a dog, looking away from the sacred event.

The group of scantily clad youths in the background contrasts dramatically with the brightly costumed multitude. An ass and an ox feed and rest contentedly in the stable; behind them horses are being unsaddled and shod. A peacock perches on the front of the stable roof, which is rendered in incorrect perspective, showing the artist's limited knowledge of this new discovery. At the side of the roof two pheasants are shown in flight. The animals and birds in the painting, though too large in scale, are realistically rendered. Blue sky, rocky hills, and ruins form the background.

It is believed that this painting was begun by Fra Angelico and finished by Fra Filippo Lippi in the fifteenth century. (*Fra* means "Friar.") The two artists used lapis lazuli and other semiprecious stones in grinding their colors. This accounts for the jewel-like quality of the hues in the painting. The exquisite colors and the superb handling of detail and composition make the painting one of the most magnificent of all time.

Fra Angelico

Fra Angelico was born near Florence, Italy, about 1400. He entered a monastery of the Dominican Order as Giovanni da Fiesole, but his gentle nature and the delicacy of his painting style soon led his fellow monks to call him Fra Angelico, or "Angelic Brother." For many years, Fra Angelico painted in the service of the monasteries in which he lived or for small neighboring churches. In 1445, however, he was called to Rome by the Pope to paint the decorations for the Chapel of Nicolas V in the Vatican, and afterward he provided paintings for churches all over Italy. Fra Angelico died in 1445 while working on a commission in Rome.

Fra Filippo Lippi

Fra Filippo Lippi was born in Florence about 1406, the son of a poor butcher. He became a Carmelite monk at the age of fifteen, and is said to have been inspired to become a painter while watching the great painter Masaccio work on the Brancacci Chapel frescoes in the Carmelite church in Florence. Fra Filippo executed paintings for the Carmelite orders in several Italian cities and was commissioned by secular leaders as well. His work was much admired by religious leaders as well as important figures like Cosimo de' Medici, who was one of his Florentine patrons. His difficult temperament often caused him problems with the Church, however, and he finally was allowed to withdraw and marry. Fra Filippo died in 1469.

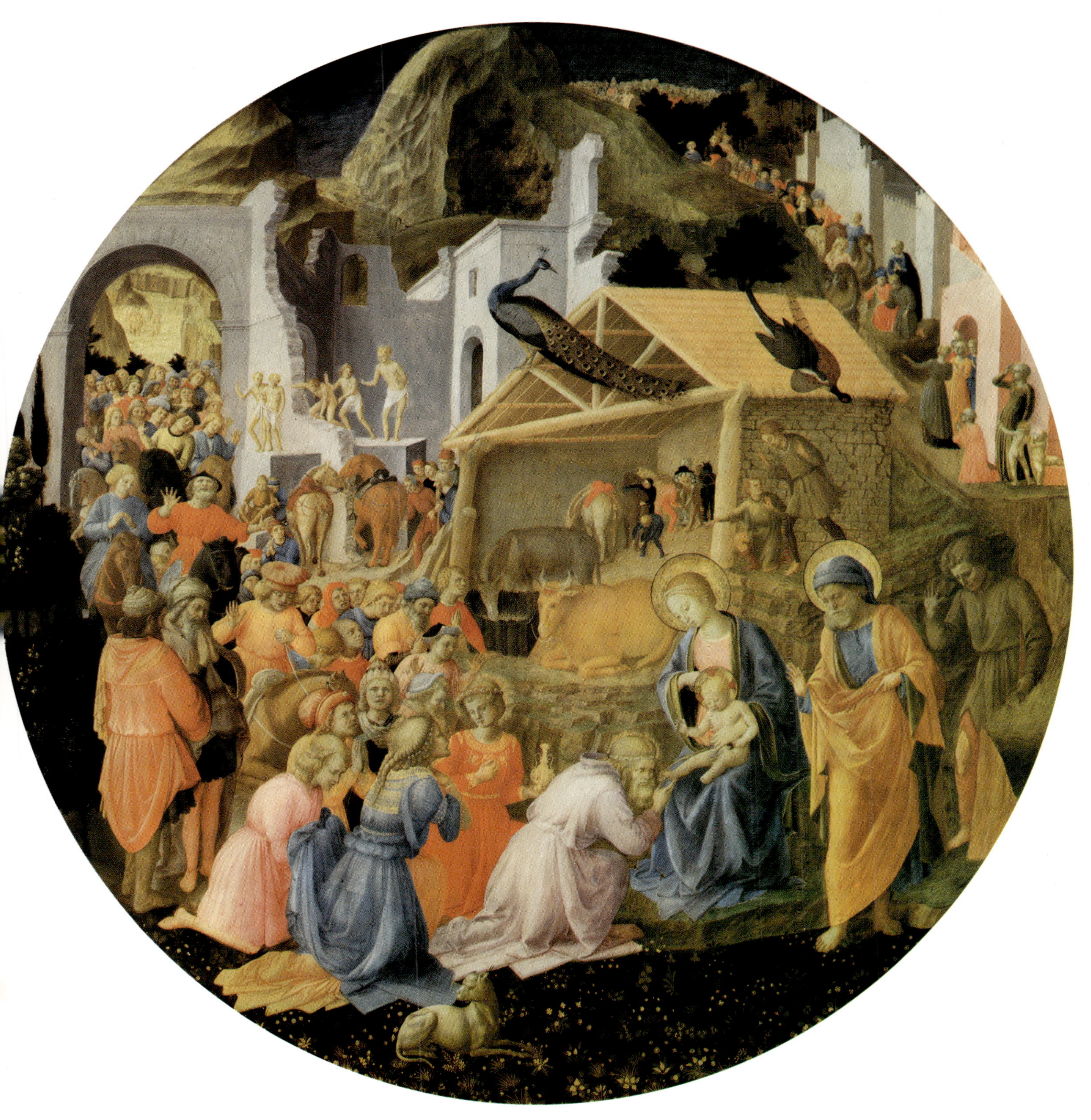

2. FRA ANGELICO and FRA FILIPPO LIPPI. *The Adoration of the Magi.* c. 1445. Wood, diameter 54″. Samuel H. Kress Collection

3. LEONARDO DA VINCI. *Ginevra de' Benci*. c. 1480. Wood, 15 1/8 × 14 1/2″. Ailsa Mellon Bruce Fund, 1967

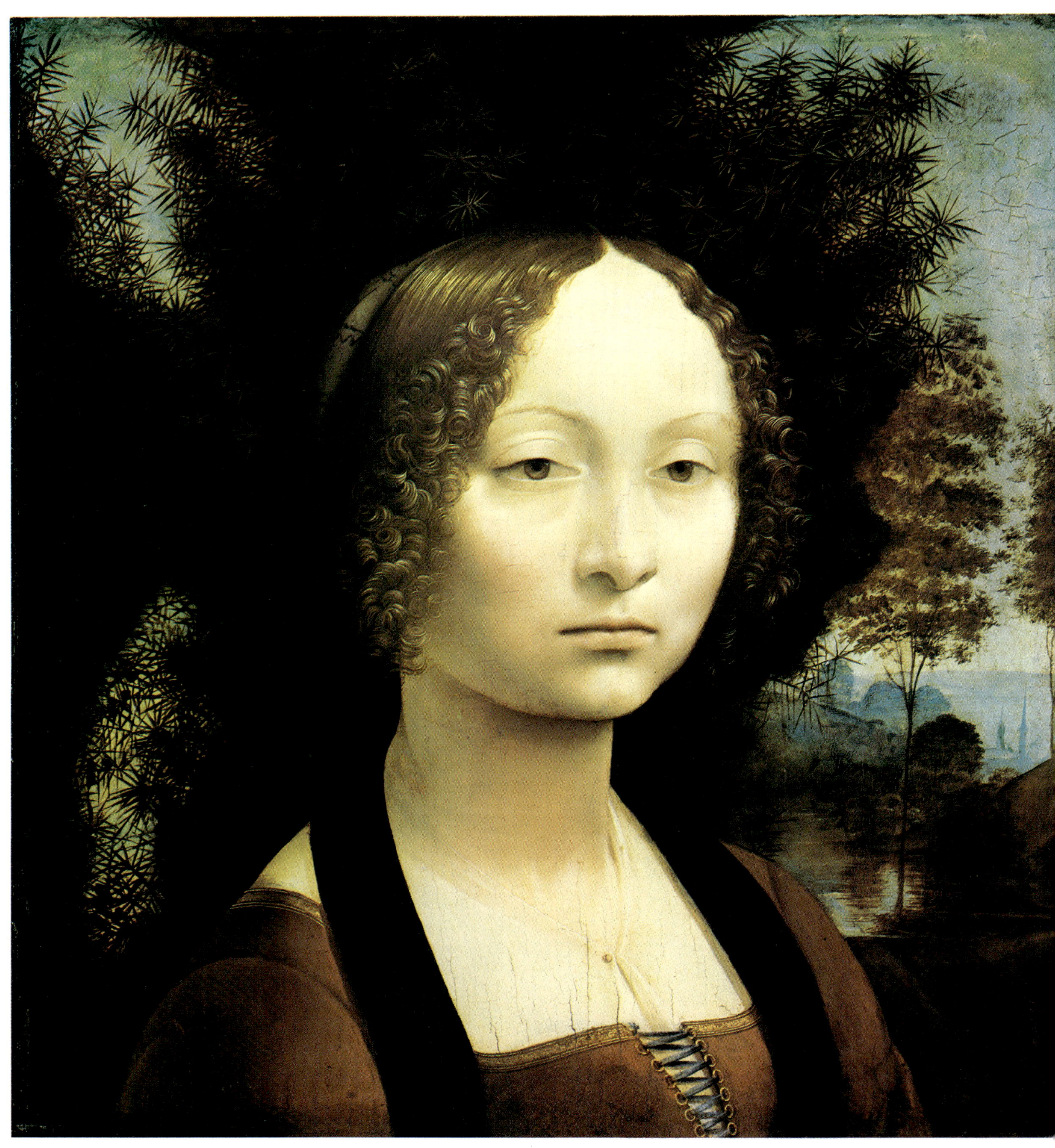

LEONARDO DA VINCI

Ginevra de'Benci

Reverse of **Ginevra de'Benci**

The portrait of Ginevra de'Benci is the only painting in America by Leonardo da Vinci, one of the greatest geniuses of all time.

Ginevra de'Benci was born in 1457, the daughter of a wealthy and powerful merchant of Florence, Italy. The family was interested in the arts and in education. Ginevra's elder brother, Giovanni, was a friend of Leonardo's; the two shared an interest in maps, rare books, and semiprecious stones.

Leonardo has placed Ginevra against a background of dark green juniper trees, which emphasizes the pallor of her calm, intelligent, yet sad face. He may have used the juniper because the name Ginevra is related to *ginepro*, the Italian word for juniper. To the right, the artist has depicted a shimmering stream, two tall trees, some shrubs, and a horizon of blue sky against which we can make out the towers, slightly blurred, of a church.

The portrait shows Ginevra in three-quarter view. The ivory flesh of her exquisitely modeled face seems to glow, and we can almost feel the softness of her skin. Her head is erect; her hazel eyes are half-closed. Her mouth is firmly closed and her lower lip protrudes slightly. Ginevra's delicately painted hair, which forms golden ringlets around her face, is covered by a small cap. The bodice of her chestnut-brown dress is edged with narrow woven gold braid and has gold rings through which blue ribbons are drawn. Her sheer white blouse is modestly fastened with a small gold button.

The reverse side of the wooden panel on which this portrait is painted pictures a spray of juniper wreathed by laurel and palm, tied together with a scroll and upon which is imprinted *Virtutem forma decorat* ("Beauty adorns virtue"). The words truly describe Ginevra de'Benci.

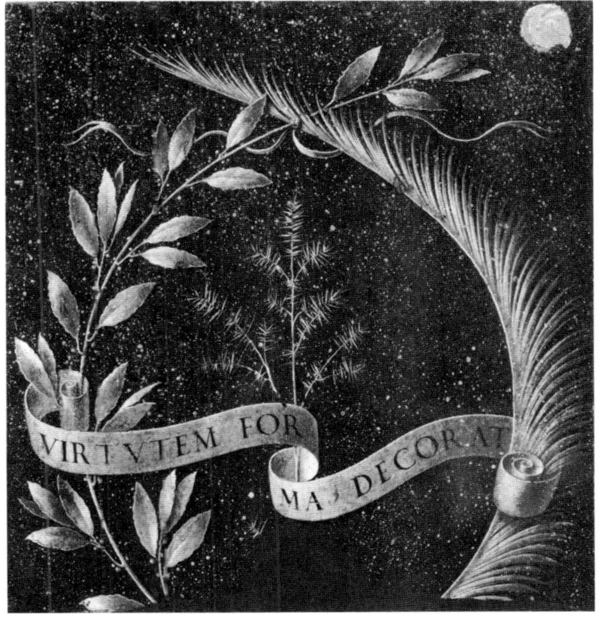

4. LEONARDO DA VINCI. Reverse of *Ginevra de'Benci*

Leonardo da Vinci

Leonardo da Vinci was born in 1452 in the Italian village of Vinci. As a child, Leonardo was intensely curious about the world and showed remarkable artistic talent. He became an apprentice to a leading artist and by the age of twenty-five had his own studio in Florence. At different times throughout his life he was supported by Lorenzo de' Medici, Ludovico Sforza (the Duke of Milan). Cesare Borgia, and King Francis I of France. The creator of such acknowledged masterpieces as the *Last Supper* and the *Mona Lisa*, he was also a distinguished sculptor, architect, scientist, and an inventor of great vision. Leonardo died in France in 1519.

5. PIERO DE COSIMO. *Allegory*. c. 1500. Wood, 22 1/8 × 17 3/8″. Samuel H. Kress Collection

PIERO DI COSIMO

Allegory

Piero di Cosimo

Piero di Cosimo was born in Florence in 1462. Like many Renaissance artists, he was interested in science. His paintings show that he understood optical perspective, or the way things look from a distance and from different angles, as well as the basics of human anatomy. He was also fascinated with Greek mythology and ancient legends, and often included references to them even in paintings of religious subjects. In addition to religious and mythological works, Piero painted portraits and designed elaborate parade floats and festival decorations as individual commissions. Toward the end of his life he became a recluse, completely withdrawing from human society. He died in 1521.

This painting by Piero di Cosimo portrays a world that never existed—a fairyland peopled with mythical figures. On a small island, a woman with wings holds a rearing horse by a string. This is, of course, impossible: a wild horse could easily break such a thin thread. In the foreground a mermaid swims through a sea strewn with plantlike forms. In the distance, birds fly over a rocky crag. The cool blue and gray colors in the sky, sea, and horse contribute to the dreamlike quality of the scene.

Against the muted sky and sea, the woman, in her intense crimson robe, wings silhouetted against the sky, seems like a vision. The tiny-headed horse, too, seems strange, prancing as it does with a dancelike step and apparently laughing.

Some scholars believe that this painting represents Aurora, the Greek goddess of the dawn, who in her horse-drawn chariot ascended to heaven each morning to announce the rising sun. If this interpretation is correct, perhaps the animal represents one of her horses; the cold blue-gray tones may indicate morning haze.

Others think that Piero was depicting "The Origin of Coral." The ancient Greeks thought that coral was seaweed that had turned to rock. They did not know that it is formed from the shells of tiny animals.

To explain how seaweed turned to stone, the Greeks invented a myth: Once there was an ugly woman named Medusa who had snakes for hair and was so frightening that anyone who looked at her turned instantly into rock. A hero, Perseus, finally killed her by a trick. Rather than looking right at Medusa, he held up his polished shield and used it as a mirror. Using the reflection to guide his sword, he cut off Medusa's head. A magic horse named Pegasus sprang from her blood. Some drops of blood fell into the ocean, causing the seaweed to become rock, or coral.

According to this interpretation, the seaweed and branches through which the mermaid swims are turning into coral. The horse may represent Pegasus, here being tamed by a nymph, a minor nature goddess whose magical powers would allow her to restrain the animal with only a string.

Since there are different ideas about the meaning of this work, you can use your imagination and form your own interpretation.

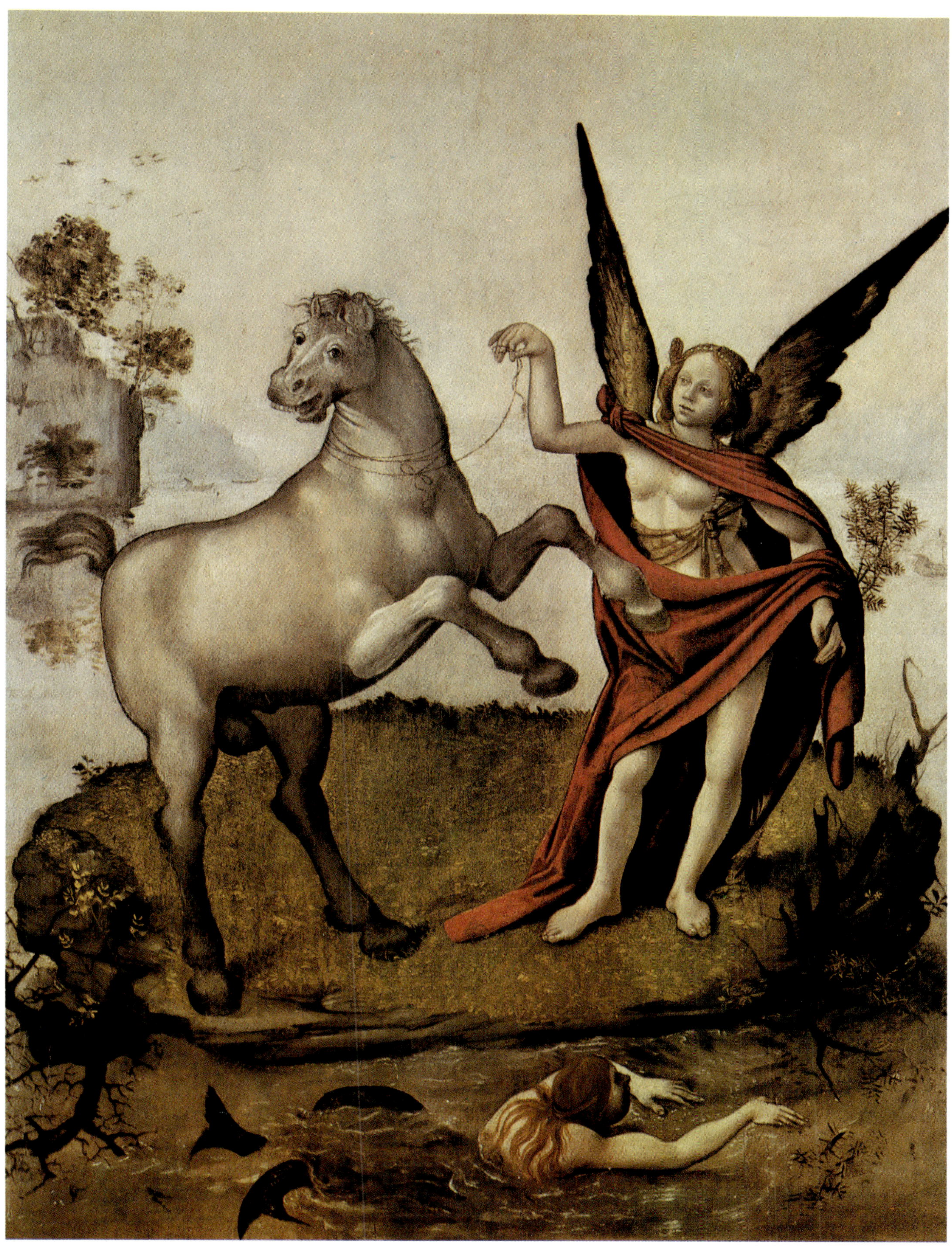

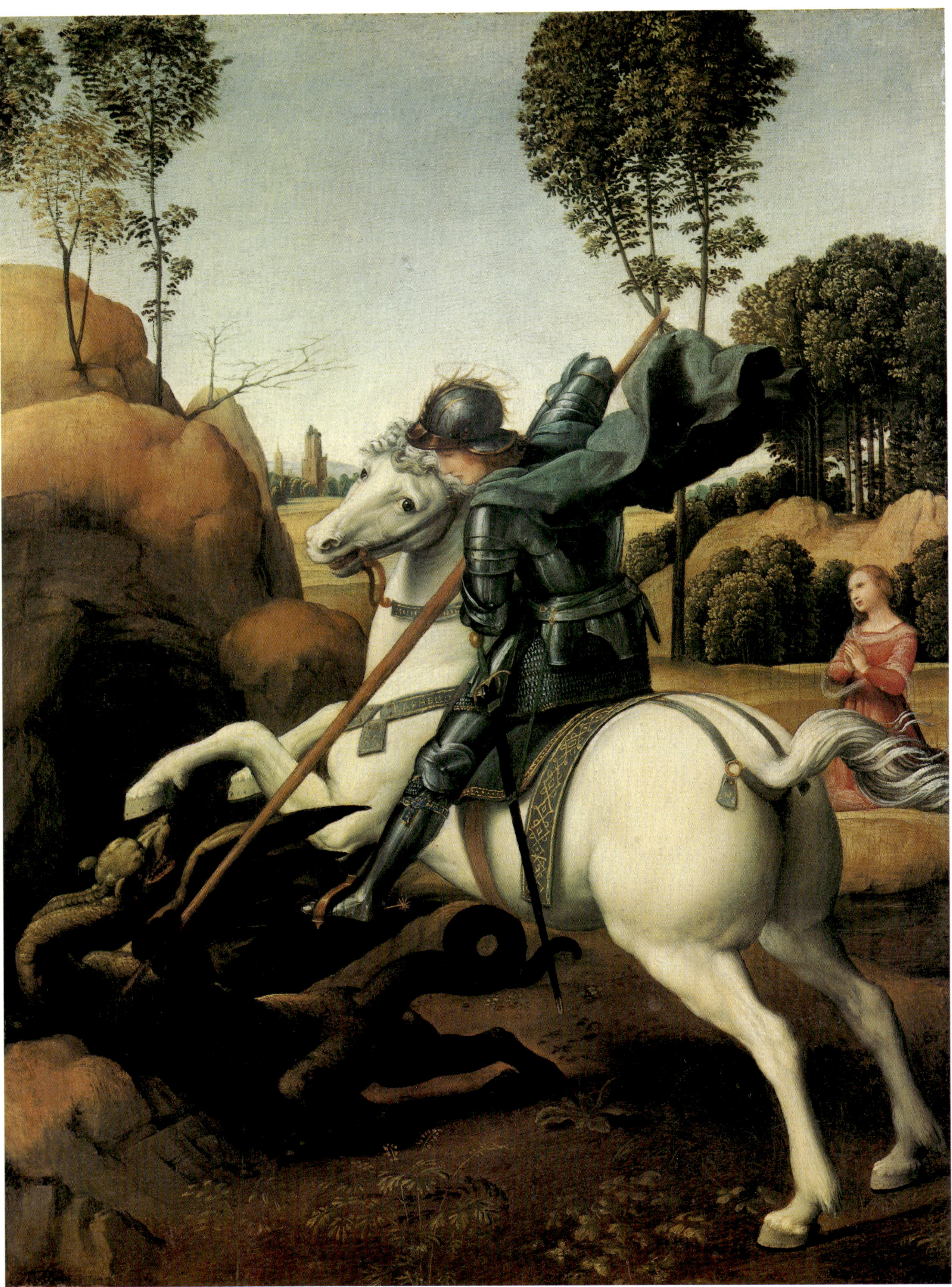

RAPHAEL

Saint George and the Dragon

ENGLISH SCHOOL

Saint George and the Dragon

The legend of Saint George and the Dragon has had wide appeal for artists over the centuries.

According to the legend, while traveling in Africa, Saint George, a Roman soldier of Christian faith, came upon a weeping princess. She was the only daughter of a pagan king, whose people were being terrorized by a dragon. To appease the beast they had fed it their sheep, then their children. Now their princess was being sacrificed.

The dragon was on the point of seizing the princess when Saint George arrived. He fearlessly galloped forward and pinned the dragon down with his lance. Tying the princess's sash around the subdued monster's neck, he led it to the city. When he saw that the people were still terrified of the dragon, he called to them to believe in God and thank Him for their deliverance. Thereupon the king and his subjects agreed to be baptized as Christians. Saint George then beheaded the dragon.

Raphael's painting was commissioned by Duke Guidobaldo da Montefeltro, the powerful ruler of the Duchy of Urbino (now a part of Italy), and was taken as a gift to Henry VII of England, who had showed his regard for the duke by making him a Knight of the Garter, England's highest honor.

Saint George wears a flowing cape and gleaming gray-blue armor. A halo encircles his head. Below his knee is a blue garter on which is inscribed *Honi* (the beginning of the motto of the Order of the Garter, *Honi Soit Qui Mal y Pense*, that is, "Evil be to him who evil thinks"). On the right the kneeling princess prays for her deliverance. Although the action portrayed in the picture is a violent one, the unmoving trees, distant towers, and light blue sky give a sense of serenity to the scene.

Saint George was the patron saint of England and the patron saint of chivalry; thus the painting was a truly fitting gift for the English monarch.

The unknown English artist who sculpted *Saint George and the Dragon* has stressed elegance, grace, and nobility in his version of Saint George and has dressed him in late-medieval armor. His shield is emblazoned with a cross. The artist has emphasized the saint's importance by making him much larger than the horse or the slender princess.

Details of the scales of the dragon, the horse's neck and tail, and the front of Saint George's leg are carved with crisp, sharp edges, while the armor plate and the horse's haunches are rounded, smoothed, and polished. The paint—in red, gold, black, and brown—was applied to emphasize certain carved portions such as the horse's bridle, the dragon's wings and tail, and the gilded edges of the princess's robe.

Although the statue's original use is not known, it has been suggested that it comes from an altarpiece.

◁ 6. RAPHAEL. *Saint George and the Dragon*. 1504/6. Wood, 11 1/8 × 8 3/8″. Andrew W. Mellon Collection

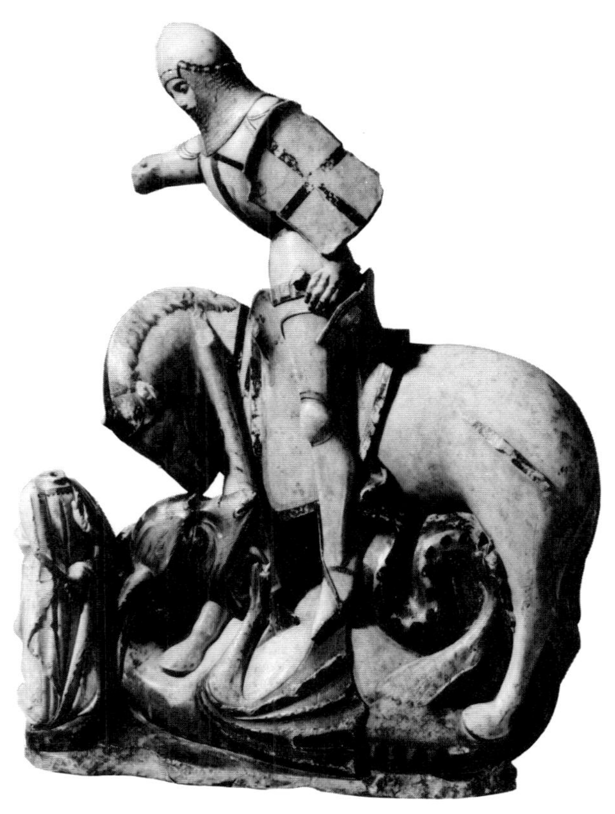

7. ENGLISH SCHOOL. *Saint George and the Dragon*. Early XIV century. Alabaster, printed and gilded, 32 × 23 3/4 × 8 1/8″. Samuel H. Kress Collection

Raphael

Raffaello Sanzio (or Santi), known as Raphael, was born in Urbino, Italy, in 1483. His father, a painter, author, and poet at the Court of Urbino, gave him painting lessons before he studied with the master Perugino in Perugia. When he went to Florence at the age of twenty-one, Raphael achieved such success that within four years the Pope invited him to Rome to decorate the *stanze* (rooms) of the Vatican apartments. Raphael became one of the most famous artists of his day, and received as many commissions as he and his assistants could satisfy. Unfortunately, Raphael's extraordinary achievements were cut short by his untimely death at the age of thirty-seven.

8. RAPHAEL. *The Small Cowper Madonna.* Probably c. 1505. Wood, 23 3/8 × 17 3/8″. Widener Collection

RAPHAEL

The Small Cowper Madonna

It was while in Florence and under the influence of the renowned artists Michelangelo and Leonardo da Vinci that Raphael painted a number of pictures of the Madonna and Child. One of these is *The Small Cowper Madonna*, probably executed when the artist was about twenty-two years old.

Seated on a stone bench, the Madonna, in a rose-red dress with a rich blue robe draped over her knees and curving over her arm, holds the young Christ Child. The Child's left leg pushes against His mother's hand while His arms encircle her neck. The expressions of both Mother and Child are tranquil and convey a sense of deep contentment. The background landscape is in shades of blue-greens, golden grays, and browns, with a gradually brightening sky. The rose tints of the flesh and the gleaming blond hair of Mother and Son contribute to the luminous radiance of the scene. The church on the right appears to be that of San Bernardino, which is near Raphael's birthplace in Urbino.

This gentle painting reflects the influence of Raphael's teacher, Perugino, in the lovely expressions of the Virgin and the Child, that of Michelangelo in the solid forms of the figures, and that of Leonardo in the subtle lights and shadows. But the remarkable talent of Raphael is revealed in his ability to combine happily the various influences and to blend them with his own unique style to produce the distinctive qualities of peacefulness, graciousness, sweetness, and ideal beauty that radiate from this painting.

The Small Cowper Madonna was so named after it came into the possession in 1780 of Lord Cowper, the British Ambassador to the Court of Tuscany, who owned another larger Virgin and Child by Raphael which is also in the National Gallery of Art.

SEE PAGE 27 FOR RAPHAEL'S BIOGRAPHY

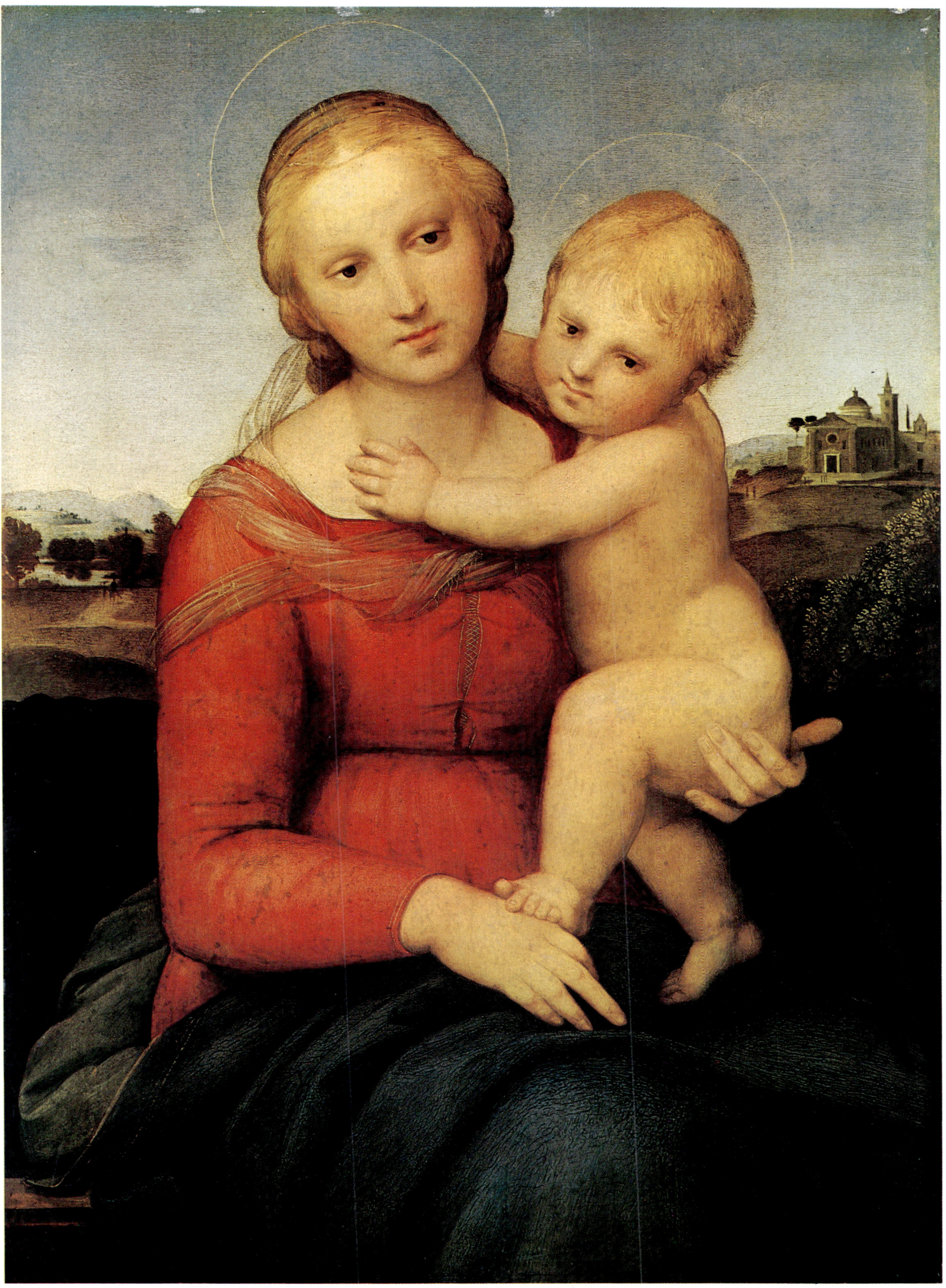

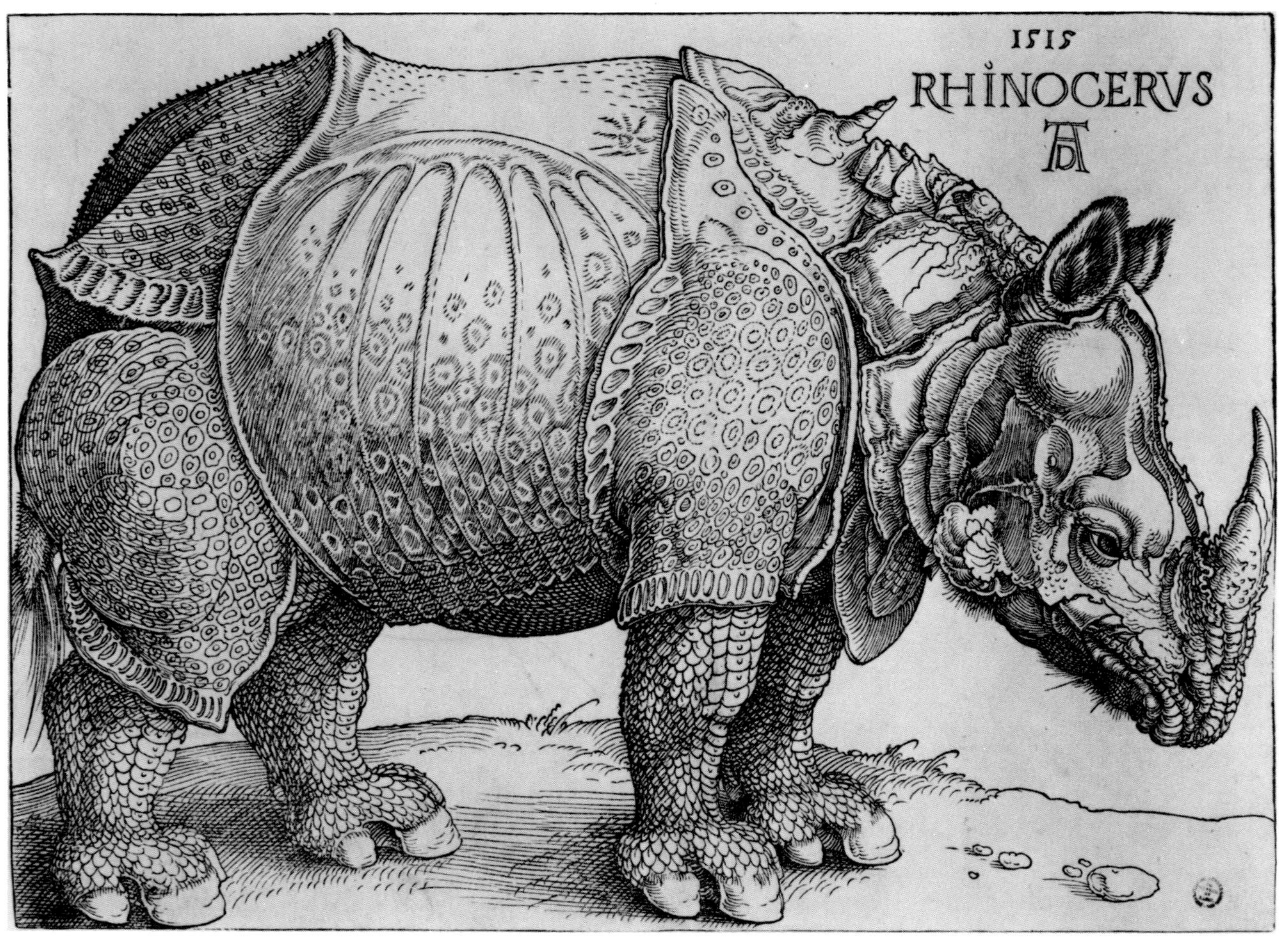

9. ALBRECHT DÜRER. *Rhinoceros*. 1515. Woodcut, 8 1/2 × 11 3/4″. Lessing J. Rosenwald Collection

ALBRECHT DÜRER

Rhinoceros

MARTIN SCHONGAUER

Elephant

Albrecht Dürer

Albrecht Dürer was born in Nuremberg, Germany, in 1471. He learned the skills of the metalworking trade from his goldsmith father before becoming a painter's apprentice. As a young man he traveled widely in northern Europe, studying painting and printmaking and making illustrations for books. After returning to Nuremberg to get married, he continued his travels, living in Italy for several years. Dürer is best known for his many engravings and woodcuts, though he also produced paintings, drawings, and decorative projects such as a prayer book for the Emperor Maximilian I, and published treatises on a variety of subjects, including geometry and military fortification. He was an acknowledged artistic and intellectual leader when he died in Nuremberg in 1528.

We see here two exotic animals, not native to Europe, as portrayed by European artists in a woodcut and an engraving. In the late fifteenth century, when Schongauer and Dürer made their prints, the woodcut and engraving processes helped spread ideas and became a type of pictorial "mass media" like movies and television in our own day.

To make the *Rhinoceros*, Albrecht Dürer began with a flat block of wood. Upon it he drew the lines which we see in the print. He then cut away the wood on either side of the drawn lines so that they were left in relief. When the design was complete, he applied ink to all the raised lines and pressed the block against a piece of paper. The raised design is what appears in black on the final print. The narrowness of the lines and the certainty with which the curves and angles are drawn attest to Dürer's great skill.

To anyone who has seen a rhinoceros, it is evident that Dürer never set eyes on the animal. He learned about the rhinoceros when the sultan of Cambay sent one to the king of Portugal as a gift. A German printer then living in Portugal saw the beast and sent to a friend in Dürer's city of Nuremberg a drawing and a description which Dürer used for his woodcut:

In the year 1513 A.D., on May 1, there was brought to Emanuel of Lisbon, the great powerful king of Portugal, such a living animal from India. They call it a rhinoceros. It is represented here in its complete form. It has the color of a speckled turtle. And it is almost entirely covered by a thick shell. And in size it is like the elephant but lower on its legs, and almost invulnerable. It has a sharp strong horn on its nose, which it starts to sharpen whenever it is near stones. The stupid animal is the mortal enemy of the elephant. The elephant fears it terribly, because where they encounter, it runs with its head down between its front legs and fatally rips open the stomach of the elephant which is unable to protect itself. Because the animal is not well armed, the elephant cannot do anything to it. They also say that the rhinoceros is fast, lively and clever.

Since Dürer had no idea of the natural habitat of the rhinoceros, he did not fill in the background. Following the description above, he shows his rhinoceros encased in hard, speckled plates which are, of course, nothing like the loose, leathery, solid gray skin of the real animal.

Dürer's contemporaries were fascinated by the fierce reputation of the rhinoceros, and the animal in the print certainly lives up to its name. As he often did, the artist dated his print and signed it with his monogram, "A.D." So that there could be no mistake, he also clearly printed the name of the animal in large letters.

Dürer's woodcut was very popular; it was printed in eight editions, and for three hundred years European artists used it as a reference when they did illustrations of the rhinoceros.

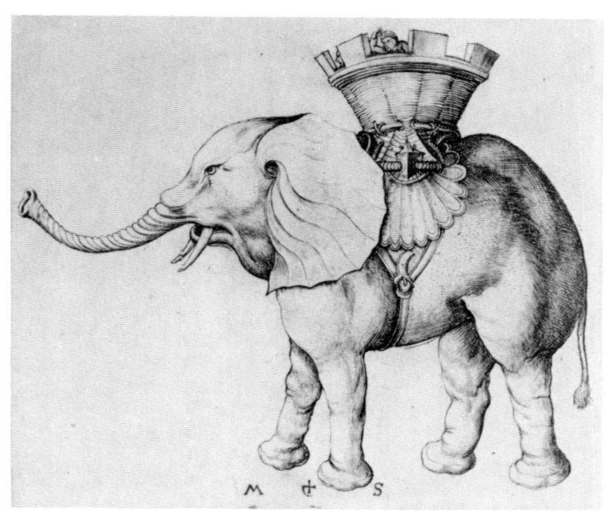

10. MARTIN SCHONGAUER. *Elephant.* c. 1480–90. Engraving, 4 1/4 × 5 3/4″. Lessing J. Rosenwald Collection

Martin Schongauer used the process of engraving for his print of an elephant. He made his engraving by incising grooves into a copper plate with a sharp tool called a burin. He then worked ink into the grooves and, after wiping clean the flat areas of the plate, used a press to force a piece of damp paper against the inked grooves. He then pulled the paper away from the plate, leaving the design imprinted on the paper. He repeated the inking and printing of the plate until he had produced the number of copies—or "impressions"—of the print that he wanted.

The elephant appears larger than life-size in contrast to the tiny men perched in the woven basket mounted on its back. The animal's trunk and ear are stylized and ornamented with shell-like swirls and curves, conveying a sense of energy. These decorative forms are repeated in the scalloped fan covering attached to the basket. The elephant's anatomy, particularly its feet, is not accurately rendered. Schongauer has given the skin a hairy quality unlike its actual thick, almost hairless appearance, and short tusks instead of the characteristic long ones. The engraving lacks the kind of detail usually found in this artist's work, another indication that it was drawn from descriptions rather than from life.

Martin Schongauer

Although he was Germany's most famous artist in the fifteenth century, little is known about Martin Schongauer's life. He was born between 1430 and 1450, and his father was a goldsmith. Though Schongauer considered himself primarily a painter, he was far more influential as an engraver. His engravings, frequently sold at fairs and religious festivals, supplied ideas for compositions and techniques to painters, printmakers, and sculptors throughout Europe. Schongauer was proud of his prints and was one of the first artists to initial the metal plates from which they were made. He died in 1491.

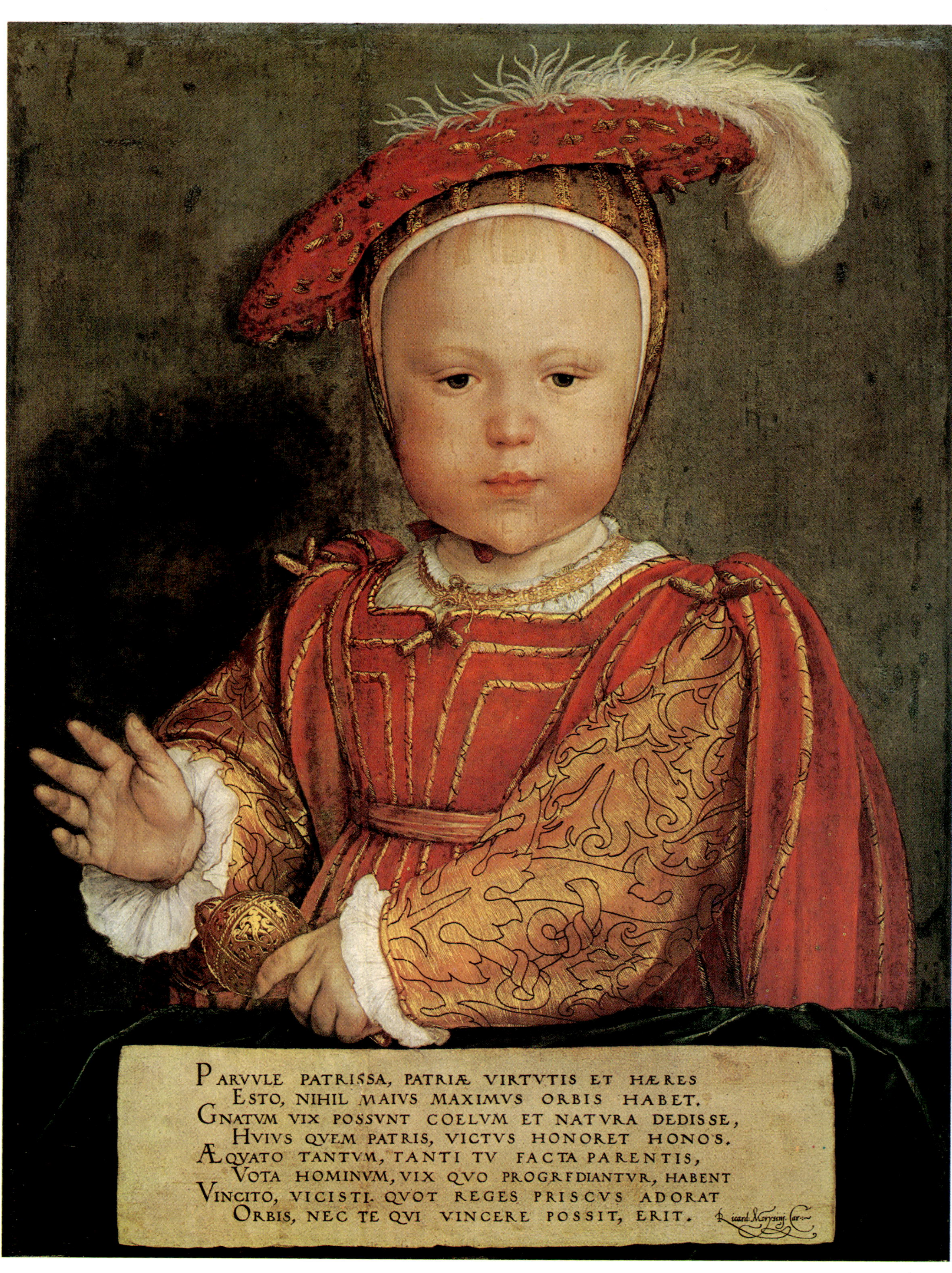
PARVVLE PATRISSA, PATRIÆ VIRTVTIS ET HÆRES
ESTO, NIHIL MAIVS MAXIMVS ORBIS HABET.
GNATVM VIX POSSVNT COELVM ET NATVRA DEDISSE,
HVIVS QVEM PATRIS, VICTVS HONORET HONOS.
ÆQVATO TANTVM, TANTI TV FACTA PARENTIS,
VOTA HOMINVM, VIX QVO PROGRFDIANTVR, HABENT
VINCITO, VICISTI. QVOT REGES PRISCVS ADORAT
ORBIS, NEC TE QVI VINCERE POSSIT, ERIT.
Ricard. Morysini. Car.

11. HANS HOLBEIN THE YOUNGER. *Edward VI as a Child.* c. 1538. Wood, 22 3/8 × 17 3/8″. Andrew W. Mellon Collection

HANS HOLBEIN THE YOUNGER

Edward VI as a Child

LUCAS CRANACH THE ELDER

A Prince of Saxony

A Princess of Saxony

It is interesting to study these portraits of royal children by Hans Holbein the Younger and Lucas Cranach the Elder. It was the fashion of the period in which these artists lived to dress the children of noble families in sophisticated, richly bejeweled adult garments—clothing which contrasted sharply with their innocent, childish features.

The portrait of the boy who became Edward VI of England was presented to his father, Henry VIII, by Hans Holbein the Younger on New Year's Day in 1539. So moved was the king by this gift that he rewarded the artist with a silver gilt cup.

It is no wonder that Henry was pleased. The portrait points up, even at Edward's early age, the marked physical likeness between father and son.

Edward, with his regal bearing, is an impressive child. He wears a miniature replica of an adult's costume. His crimson doublet has slashed sleeves; his bodice and tunic are trimmed with gold brocade. White ruffs are at his wrists, and gold embroidery encircles the neckline of his shirt. The red velvet bonnet with its embroidery and ostrich plumes perches gaily on his close-fitting coif. Edward's right hand is raised in a kingly gesture; in his left hand he holds a gold rattle.

Love and hope are conveyed in the portrait's Latin inscription: "Little one, emulate thy father and be the heir of his virtue; the world contains nothing greater. Heaven and earth could scarcely produce a son whose glory would surpass that of such a father. Do thou but equal the deeds of thy parent and men can ask no more. Shouldst thou surpass him, thou hast outstript all kings the world has revered in ages past." This flattering counsel was in vain. In his sixteenth year, six years after he had become king, Edward VI died.

Hans Holbein the Younger

Hans Holbein the Younger was born in Germany in 1497. His father, a painter, was his first art teacher. At eighteen, young Holbein went to Basel, Switzerland, where he began his career as a professional painter. He was drawn on two separate occasions to London in hopes of achieving fame and fortune. He painted Sir Thomas More and other important personages and eventually became court painter to Henry VIII. At the height of his long-awaited success, Holbein fell victim to the plague and died in London in 1543.

Lucas Cranach the Elder

Lucas Cranach the Elder was born in Germany in 1472 and studied engraving with his father. One of his sons, Lucas Cranach the Younger, became his assistant. As court painter to Frederick the Wise in Wittenberg, Cranach painted, made woodcuts and engravings, and designed dies for the mint. In return, Frederick gave Cranach not only a printer's patent and the exclusive right to print Bibles, but also a monopoly on the sale of medicines in Wittenberg. After Frederick's death, Cranach served two of his successors until his own death in 1553.

The portraits of the Saxon prince and princess shown here were painted by Cranach when he was court painter in Wittenberg, Germany. It is believed that they represent the children of Duke George the Bearded of Saxony. In their poses and expressions they seem more like real, flesh-and-blood children than does Holbein's idealized prince. There is a fascination about the boy. His wavy blond hair is subtly varied by lights and shadows. The delicate wreath he wears is fashioned of golden wire, pearls, and precious stones, clearly painted with Cranach's customary attention to detail. Most entrancing, however, are his eyes, which are fixed on the observer. The young prince's left hand clutches the lapel of his outer robe uncertainly, as if he is not yet accustomed to the feel and proper set of his princely garments.

As in the painting of the prince, a very dark background is used to set off the delicate face of the gentle princess. Her wavy blond hair reaches to her waist. Her richly brocaded dress is accented by glittering heavy gold chains that form a yoke and blend with her hair. She also wears an elaborate gold necklace. Her arms are crossed and her fingers are curled, and there is a somewhat wistful, faraway look in her eyes.

12. LUCAS CRANACH THE ELDER. *A Prince of Saxony*. c. 1517. Wood, 17 1/8 × 13 1/2″. Ralph and Mary Booth Collection

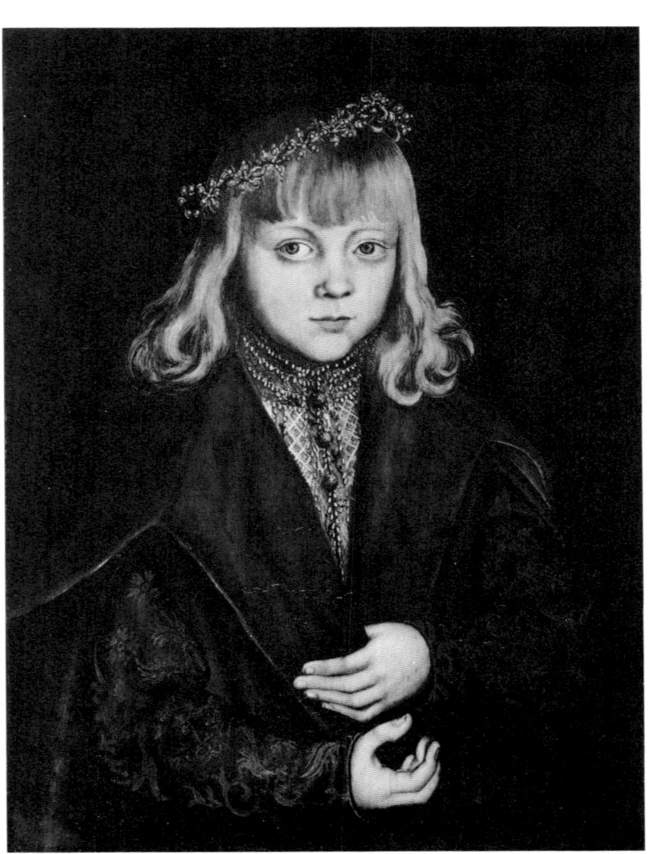

13. LUCAS CRANACH THE ELDER. *A Princess of Saxony*. c. 1517. Wood, 17 1/8 × 13 1/2″. Ralph and Mary Booth Collection

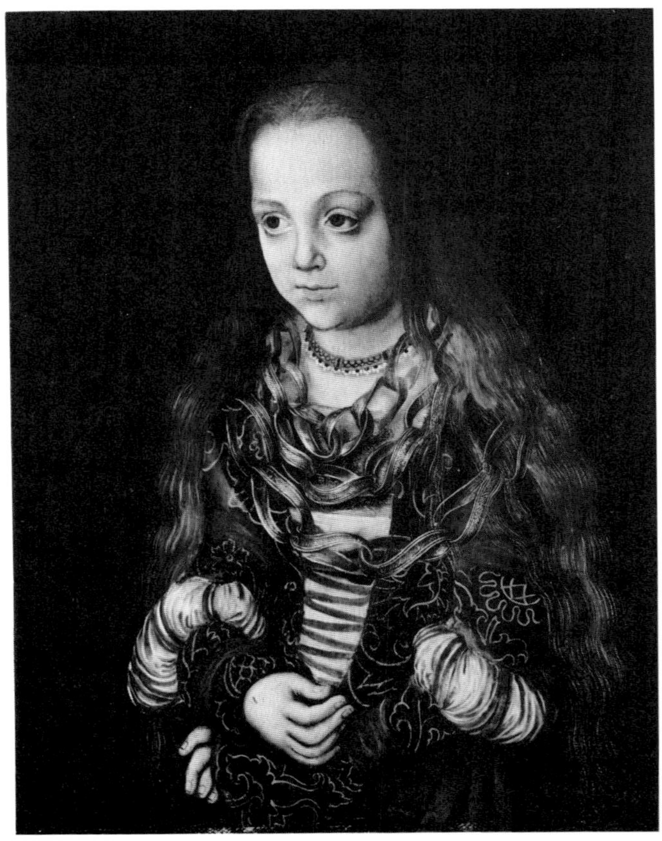

DESIDERIO DA SETTIGNANO

Bust of a Little Boy

PIERRE LEGROS

Cherubs Playing with a Lyre

JEAN-BAPTISTE TUBI

Cherubs Playing with a Swan

It would be hard to find two more different sculptural styles in depicting children than the Desiderio *Bust of a Little Boy* and (overleaf) the fountains by Tubi and Legros. The Desiderio reveals certain individual peculiarities. The fountains' cupids, in contrast, are ideal types, not individuals.

The Renaissance sculptors of the 1400s wanted their work to express a quiet dignity: the Desiderio child has only a slight tilt of the neck to indicate the boy's curiosity. In the 1700s, however, when the fountains were made, the taste was for robust energy and action in sculpture. The cherubs with the swan look so active that you can imagine them falling out of their seashell at any moment. And even the cherubs sitting with the harp adopt energetic postures and active poses which make the composition lively and exciting.

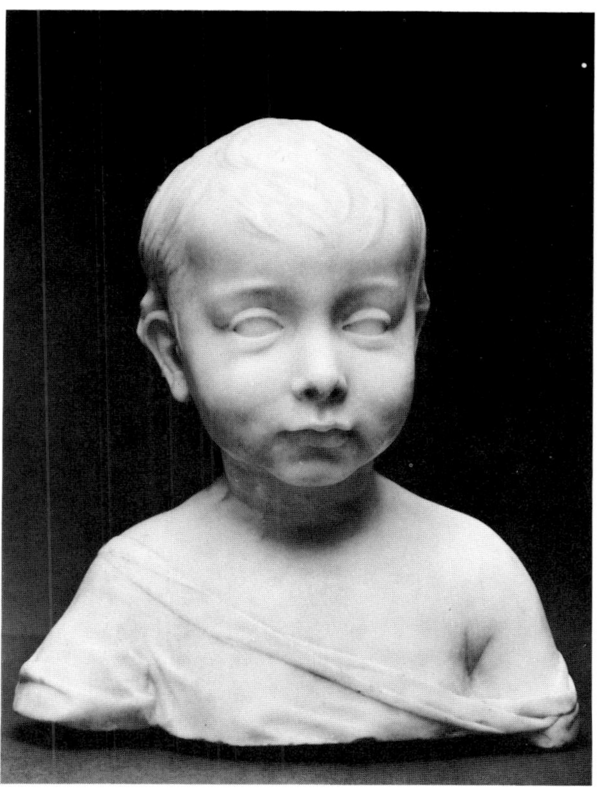

14. DESIDERIO DA SETTIGNANO. *Bust of a Little Boy*. c. 1453–63. Marble, 12 × 10 3/8 × 6 3/8″. Samuel H. Kress Collection

Although the little boy portrayed in Desiderio's sculpture is unknown, he was probably the son of some wealthy Florentine merchant. Children's portraits in the Renaissance are thought to have had a double purpose—also serving as images of religious figures such as the Christ Child or the infant Saint John the Baptist. This would seem to be the case with this carved marble bust: there is a small hole at the top of the head where a metal halo may once have been attached. These portrait busts may have been placed on altars in the private chapels of Florentine families.

Few subjects are more difficult to portray than very young children: they have not yet developed individual bone structures and the stamp of personality is not yet clearly imprinted on their faces. Yet Desiderio succeeded in capturing the fresh charm of this little boy with great technical skill. The sculptor truthfully recorded the specific features of a particular child—one with very wide-set eyes and full cheeks. His ears, like those of many children, stick out slightly. This child seems almost ready to break into a burst of youthful laughter.

Desiderio also has conveyed the softness that is so much a part of a baby's lovableness. The hard stone seems warm and living. We feel that the child's cheek might respond to the pressure of a touch, and we want to reach out and rearrange his drapery.

Desiderio carefully polished the face and shoulders to create

Desiderio da Settignano

Desiderio da Settignano was born near Florence about 1430. He is believed to have studied under the great and influential Florentine sculptor Donatello, and was already established as an independent sculptor by the age of twenty-five. He is recognized as one of the finest sculptors of children who ever lived, and some of his most charming works are enhanced by cherubs and mythological cupids. His portraits of specific children seem to have been an important source of his livelihood. The artist died in his early thirties, in 1464.

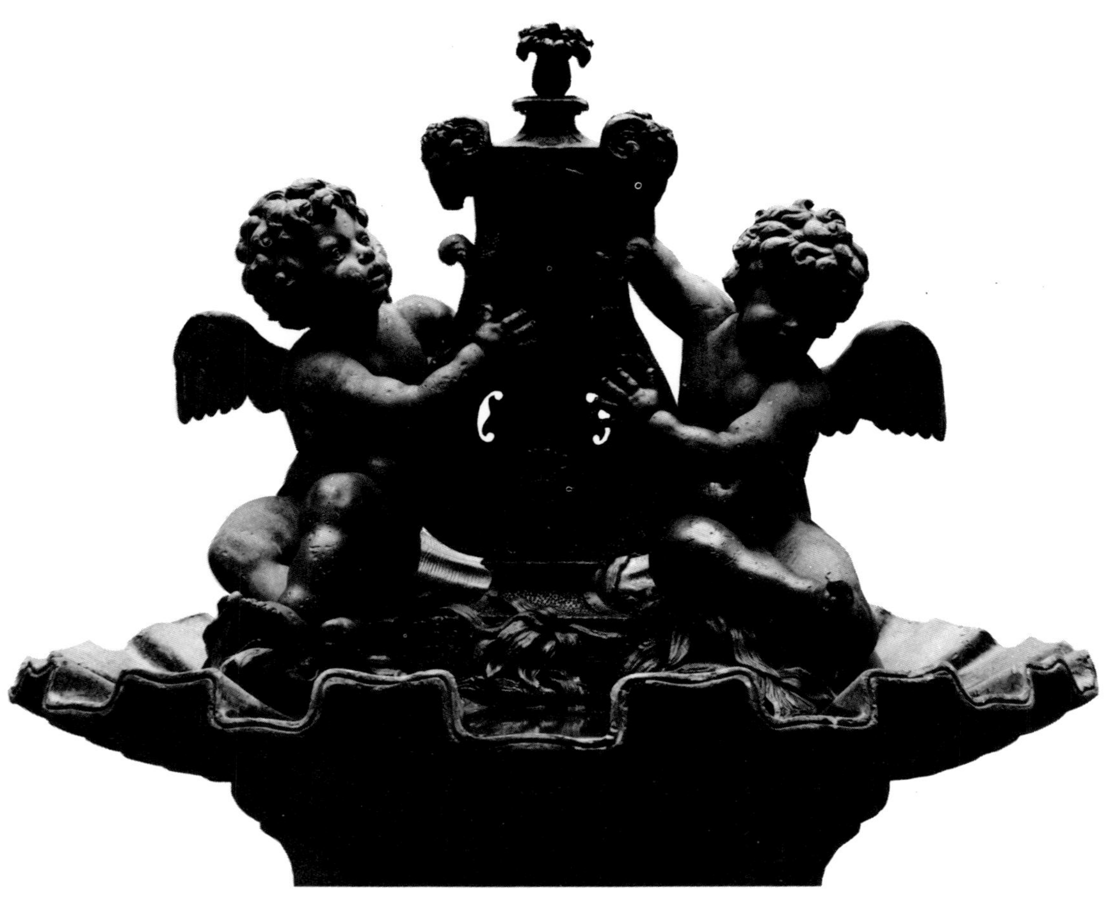

15. PIERRE LEGROS. *Cherubs Playing with a Lyre.* Cast c. 1670–75. Lead, traces of gilding, 43 × 59″. Andrew W. Mellon Collection

the effect of a child's fine, velvety skin. Contrasting with the smooth skin are rougher textures. Neither the flannel-like drapery nor the downy hair is as highly polished as the flesh.

The fountains by Legros and Tubi are the only two surviving pieces from the celebrated Théâtre de l'Eau, or "Water Theater," at Versailles, an outdoor arena where grassy terraces cut into the lawn enabled the aristocracy to watch performances of fountains that could shoot, cascade, trickle, bubble, or spray the water piped to them. Both are working pieces.

The theme of these two sculptures, as with all the others in the Water Theater, is love.

In Legros's work, now in the National Gallery's West Garden Court, two cherubs, or little angels, hold up a lyre. This ornate harp, decorated with rams' heads, is one of the musical instruments which was used to accompany love songs. Another name for cherubs is "cupids," and winged babies playing lyres often appear on Valentine's Day cards. Resting in the seaweed are quivers of arrows. An arrow shot from a cupid's bow causes an individual to fall in love with the next person he sees. The base of the fountain is a giant seashell, yet another symbol of love. Venus, the goddess of love and beauty, emerged from the sea and was carried to shore in a seashell.

In Tubi's fountain, which is in the Gallery's East Court, cherubs

Pierre Legros

Pierre Legros was born near Paris in 1629. He studied with a court sculptor before becoming a member of the Royal Academy of Painting and Sculpture in Paris. In addition to the statues for the gardens at Versailles, Legros received many other commissions from Louis XIV, including a carved relief panel project for Les Invalides, an enormous church being built by the king, and decorations for the Palace of the Louvre. During his later years, Legros devoted much of his time to teaching. He died in Paris in 1714.

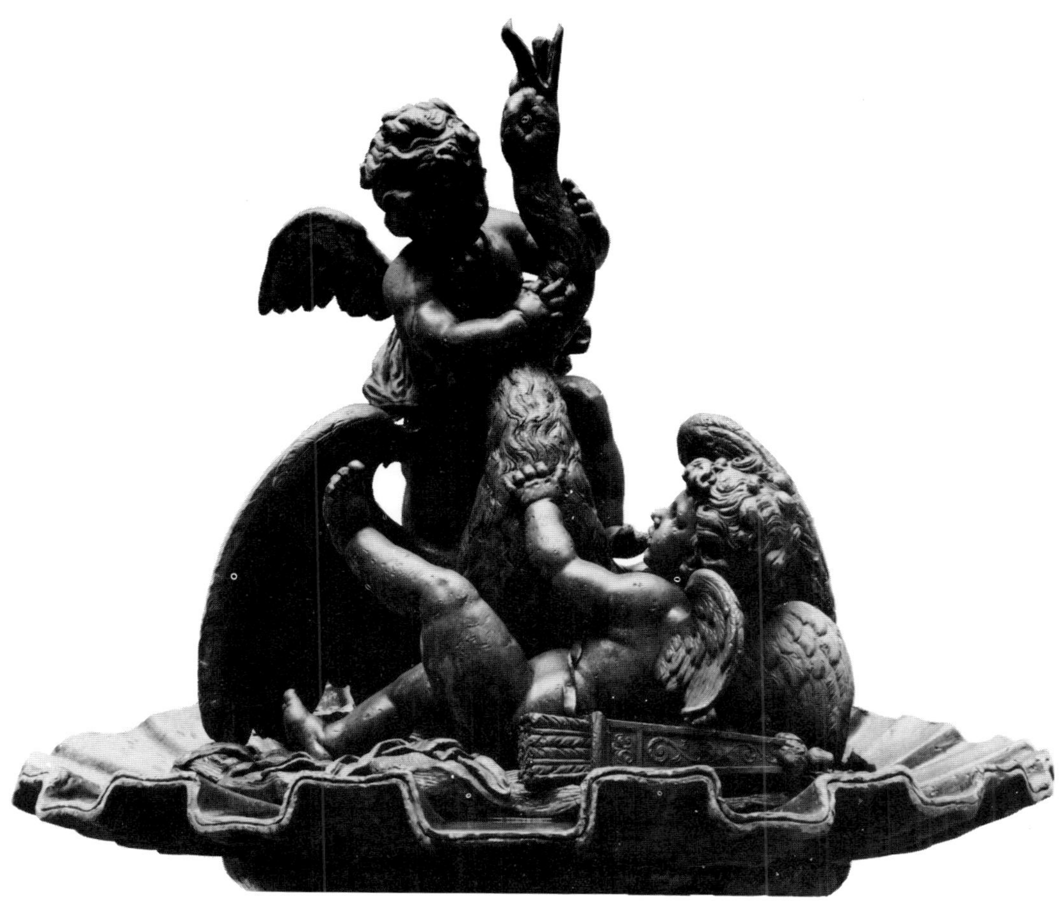

16. JEAN-BAPTISTE TUBI. *Cherubs Playing with a Swan.* Cast c. 1674. Lead, traces of gilding, 47 × 59″. Andrew W. Mellon Collection

frolic with a swan, the graceful bird that is another of the many symbols of love. The swan's beak is the source of the water; the pipe can be seen in the bird's mouth. Like Legros's cupids, Tubi's winged cherubs are delightfully chubby, with full cheeks and fat toes and fingers.

The seashell basin of the Tubi fountain is so nearly identical to the basin of Legros's piece that it is assumed they were cast from the same mold. On close inspection, though, one can detect slight differences in the two fountains. The group by Legros is more compact in design. Also, Legros's sculpture, with a cherub on either side of the lyre, is more formal and dignified than the Tubi, where one cherub playfully climbs on the swan's back while the other topples backward into the shell.

The fountains are cast of lead, a metal that doesn't rust—even under constantly running water. Lead's natural color, however, is an unattractive dull gray. If you look carefully at the photographs, you can see traces of the original gold leaf that was used to gild the fountains.

Since the audience had to view the fountains from a considerable distance, the sculptors exaggerated their shapes. The textures of the cherubs' hair and wings and the swan's feathers are coarser than in nature. With the surfaces so bold, the designs could be seen clearly from the arena. Also, the figures are much larger than life-size; the "babies" are nearly four feet tall!

Jean-Baptiste Tubi

The sculptor Jean-Baptiste Tubi was born in Rome in 1635 and was trained in Italy. When he was about twenty years old, however, he moved to France and lived in Paris for the rest of his life. Like Legros, Tubi became a member of the French Royal Academy of Painting and Sculpture, and through his friendship with Charles Le Brun, chief court painter to Louis XIV, and Antoine Coysevox, the king's favorite sculptor, he received many important commissions. A teacher as well as practicing sculptor, Tubi inspired many young artists. He died in 1700.

EL GRECO (DOMENIKOS THEOTOKOPOULOS)

Saint Martin and the Beggar

Saint Martin was born about A.D. 316, in the area that is now Hungary, during the reign of Constantine the Great. He was forced to join the imperial cavalry by his father, a military tribune, and was stationed in France at Amiens. One bitter day at the gates of the city he met a naked beggar. Taking pity on the poor shivering man, Martin cut his cloak in half and shared it with him. That night Christ, wearing the halved cloak over his shoulders, appeared to Martin in a vision, saying, "What thou hast done for the poor man, thou hast done for me." Martin's experience left him so greatly moved that he converted to Christianity and eventually left the army.

The languid, elongated figures, shimmering lights, and brooding shadows in El Greco's painting are uniquely expressive of the intensely religious spirit of Spain in the sixteenth century. Saint Martin is seated astride a white charger. The simple black harness accentuates both the whiteness of the horse and its great size. Curly blond hair frames the saint's gentle and compassionate face. He is dressed in heavy gleaming metal armor richly decorated with gold. By his side, the beggar grasps the yellow-green cloak, which partially covers him. There is a mystical and emotional quality about the sinuous figure of the beggar which we feel immediately. The intense blue sky and white clouds contribute to the unearthly brilliance of the picture. The city of Toledo, Spain, where El Greco lived for much of his life, is seen in the distance between the firmly planted leg and raised hoof of the horse.

El Greco has forcefully conveyed the message of charity and conversion in the painting, one of three done between 1597 and 1599 as altarpieces for the Chapel of San José in Toledo.

El Greco (Domenikos Theotokopoulos)

Domenikos Theotokopoulos became known as "El Greco"—which means "The Greek" in Spanish—because, although he spent much of his life in Spain, he was born in 1540 on the Greek island of Crete, a Venetian territory at that time. After some training in the Byzantine tradition of his birthplace, he studied with Titian, the great Italian master, in Venice. Eventually he settled in Toledo, Spain, where there was more enthusiasm for the religious fervor and drama of his paintings than he had found in Italy. He was able to live in splendor in his adopted city, and he died there in 1614.

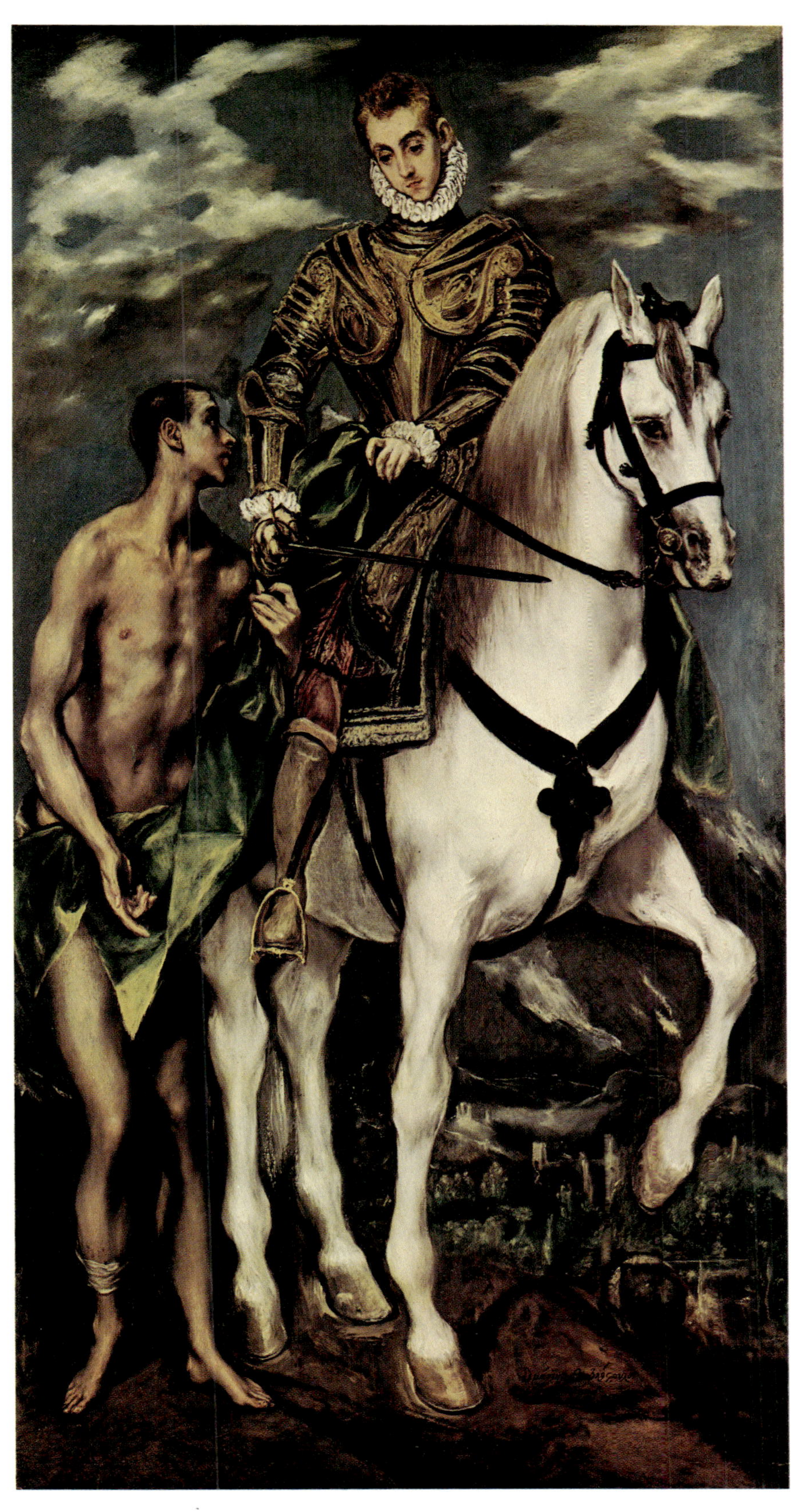

17. EL GRECO (DOMENIKOS THEOTOKOPOULOS). *Saint Martin and the Beggar*. 1597/99. Canvas, 76 1/8 × 40 1/2″. Widener Collection

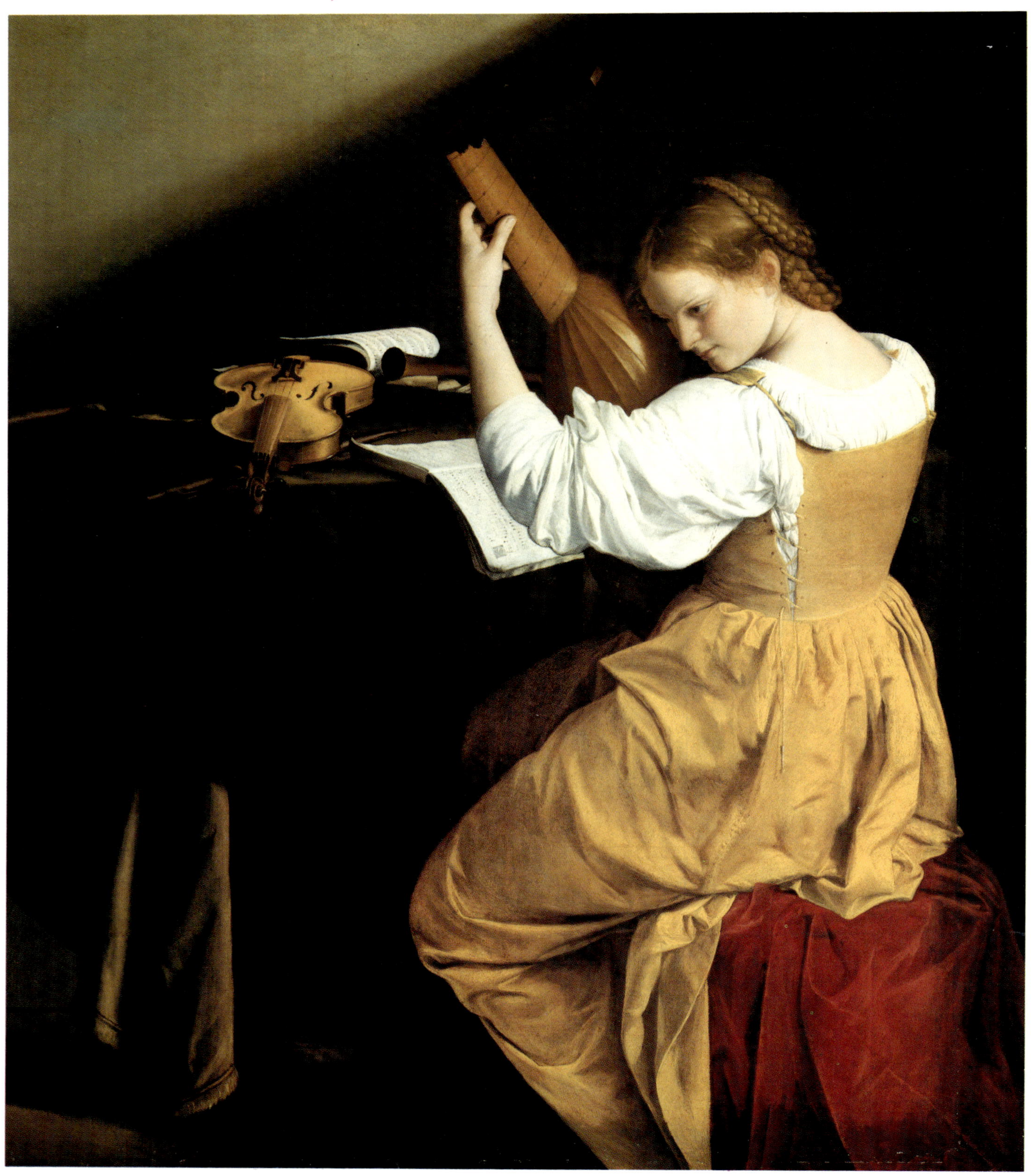

Orazio Gentileschi

Orazio Lomi, who adopted his mother's maiden name of Gentileschi, was born in Pisa, Italy, in 1563. He studied painting first with his brother Aurelio and later in Rome. After working in Rome, Genoa, and Turin, he went to Paris, where Marie de' Medici was his patron. Having established his reputation as a painter, Gentileschi then traveled to London, where he became a naturalized citizen and executed works for many members of the English nobility, including King Charles I. Gentileschi died in London in 1639.

18. ORAZIO GENTILESCHI. *The Lute Player.* Probably c. 1610. Canvas, 56 1/2 × 50 5/8″. Ailsa Mellon Bruce Fund

ORAZIO GENTILESCHI

The Lute Player

JEAN-BAPTISTE-CAMILLE COROT

Gypsy Girl with Mandolin

Orazio Gentileschi has so beautifully highlighted *The Lute Player* that we are attracted to her at once.

The young girl, who was probably Artemisia Gentileschi, the artist's daughter, is seated on a stool draped with rich red textured cloth, before a table covered in green velvet. Her blond hair is done up in a double looped braid. Her face is gentle, and her delicate skin has a lovely whiteness. The folds of her golden dress are skillfully depicted. On the table, which blends into the dark background, we can discern two open books of musical scores, a violin and a bow, and three recorders (old wind instruments somewhat like pipes or flutes). The warm brown wood of the lute, a seventeenth-century precursor of the guitar, glows against the background and brings out the whiteness of the girl's fingers as she tunes the instrument. Her lovely head is bent and her expression is concentrated as she listens for the right pitch. Perhaps she is waiting for others to join her in a quartet.

So realistically has Gentileschi created this picture that we seem to hear the plucked notes. We can easily imagine the feel of the fine textures of the cloth, of the girl's soft skin and hair, and of the warm smooth woods of the instruments.

Artemisia Gentileschi was a person of many talents. She was to become one of the most eminent woman painters of her time.

The young girl in Corot's *Gypsy Girl with Mandolin* has a pensive expression—she seems lost in her dreams as she strums her mandolin.

Corot has placed the gypsy against a forest background which is softened and blurred, creating a poetic mood in the painting. The girl is robed in a loose dress with a rose at her bosom. Her flowing dark hair, partially covered by a cloth, frames her exotic face. Perhaps she is remembering a lost sweetheart as she plays the sweet sad music.

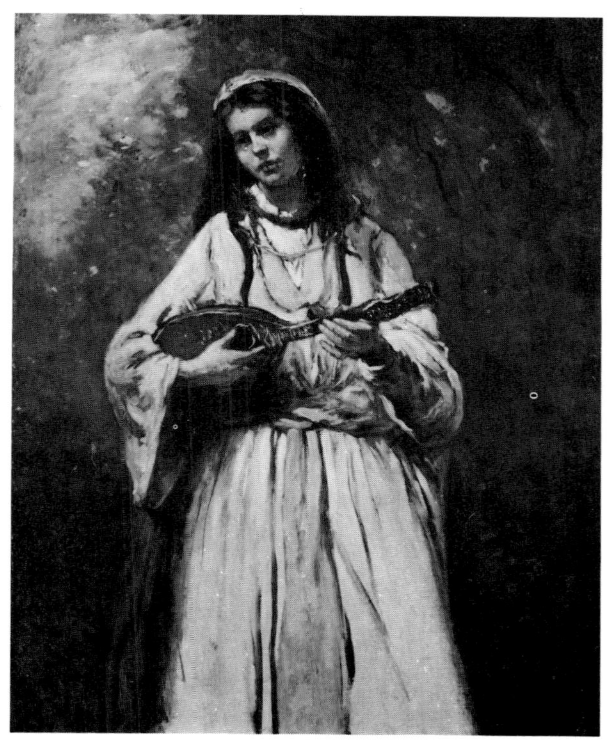

19. JEAN-BAPTISTE-CAMILLE COROT. *Gypsy Girl with Mandolin.* Probably c. 1870/75. Canvas, 25 5/8 × 21 5/8″. Gift of Count Cecil Pecci-Blunt

Jean-Baptiste-Camille Corot

Jean-Baptiste-Camille Corot, the son of a successful Parisian milliner, was born in 1796 in Paris. When Corot was twenty-six, his father finally consented to his son's desire for an artistic career. After much study, travel, and hard work, Corot won acclaim as a landscapist, and in 1846 he was decorated with the Legion of Honor. Later in life he perfected the luminous dream-landscape style for which he is best known, and stimulated interest in painting nature in the open air instead of from memory in the studio. Some of Corot's youthful figure studies, though practically ignored during his lifetime, are today considered to be as interesting as his landscapes. Corot died in Paris in 1875.

20. JAN DAVIDSZ. DE HEEM. *A Vase of Flowers.* c. 1645. Canvas, 27 3/8 × 22 1/4″. Andrew W. Mellon Fund

JAN DAVIDSZ. DE HEEM

A Vase of Flowers

In *A Vase of Flowers,* Jan Davidsz. de Heem has challenged the observer to discover a number of delightful surprises. There are more than twenty different varieties of flowers—red and white tulips, pink roses, a red carnation, white candytuft, blue morning glories, coral sweet peas, and a white poppy are but a few. All these plants could have been found growing in a seventeenth-century Dutch garden, although they would not all have been in bloom at the same time.

De Heem's flowers were probably painted both from nature and from botanical illustrations. He has given the lavish bouquet a charming casual arrangement, yet has painted each form with exact precision. In the deep green glass vase we can see colorful stems and a clearly drawn water line. There is also a reflection of a window, the bit of sky serving to remind us of the outside world. Notice the small snail on the edge of the marble table in the lower-left corner; nearby are a lizard and a spider. Also to be seen are an ant, a moth, butterflies, a caterpillar, and a bee. Each of these creatures could serve as a complete painting in itself.

In De Heem's time, Dutch art reflected the interest of the Dutch people in symbols. The lizard, moth, and spider in the painting signify night—an active time for these creatures; the honeybee indicates thrift and industry; the caterpillar and butterfly convey thoughts of death and rebirth. For us today, *A Vase of Flowers* remains a joy, filled with intriguing surprises.

Jan Davidsz. de Heem

Born in Utrecht, the Netherlands, in 1606, Jan Davidsz. de Heem studied painting with his father. He ultimately settled in Antwerp, where his work was much admired by Flemish painters and collectors. De Heem specialized in still-life compositions, especially flowers and fruits, which were extremely popular during the seventeenth century. His works are marked by their luxurious, decorative quality. He died about 1684.

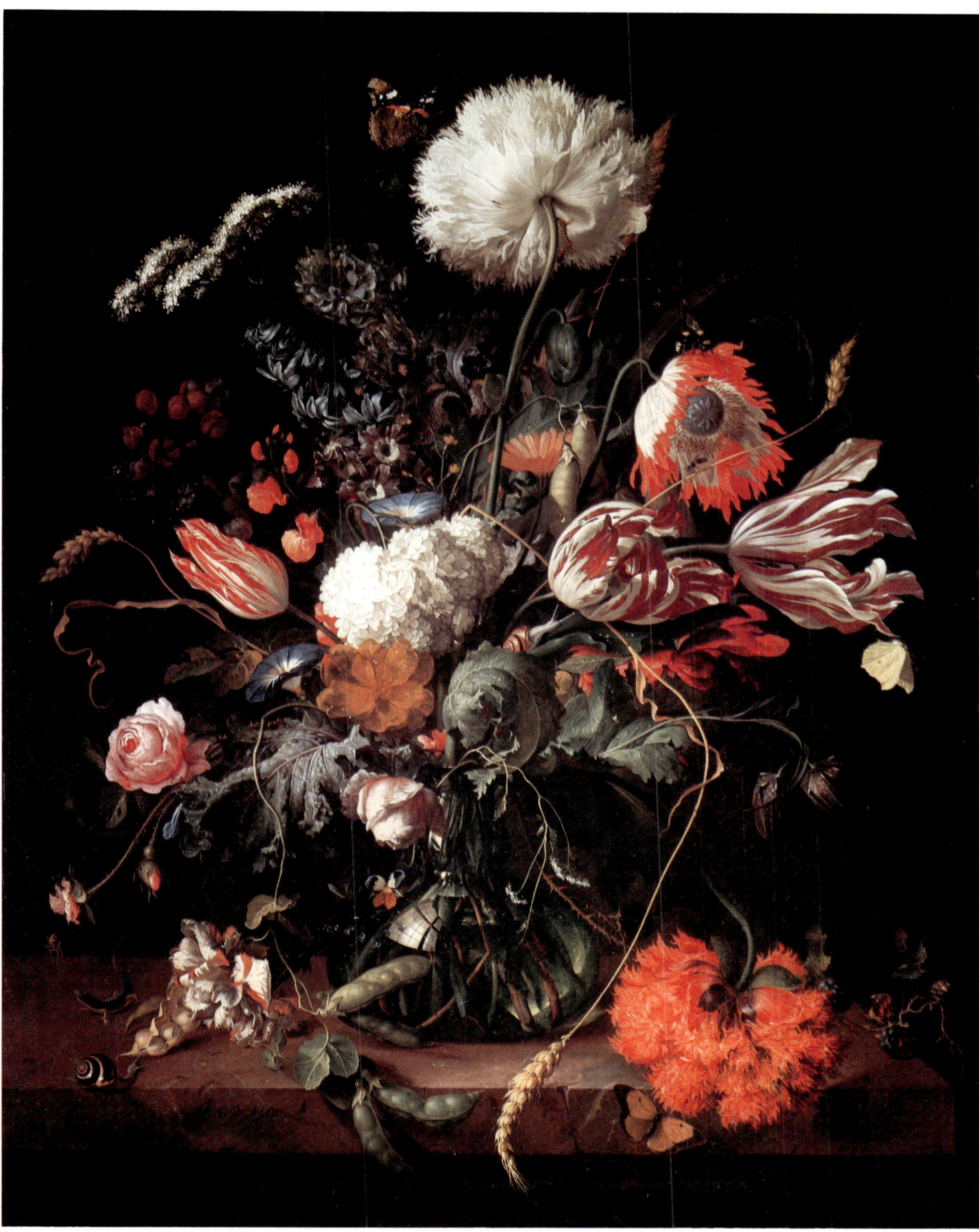

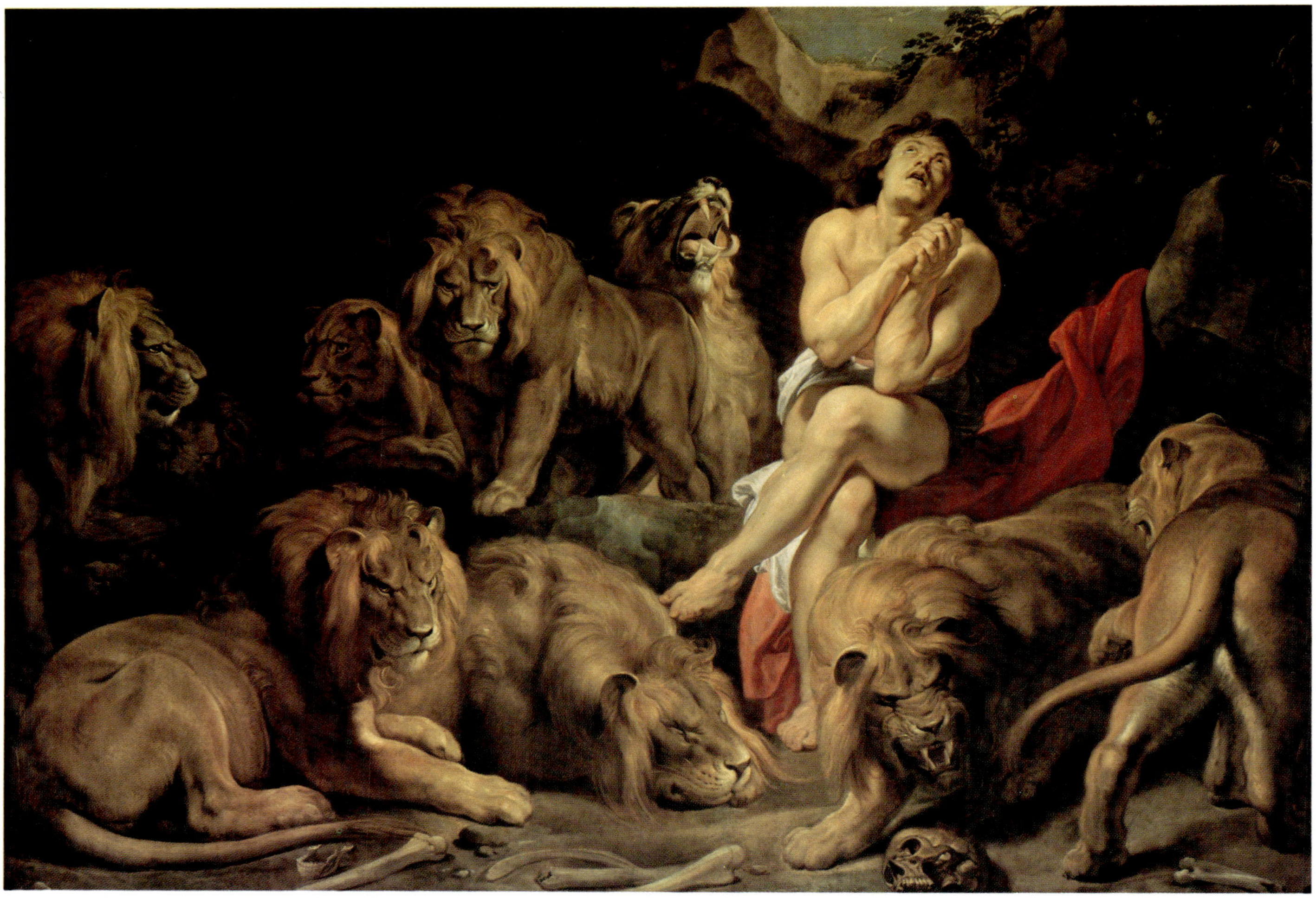

21. PETER PAUL RUBENS. *Daniel in the Lions' Den.* c. 1615. Canvas, 88 1/4 × 130 1/8″. Ailsa Mellon Bruce Fund

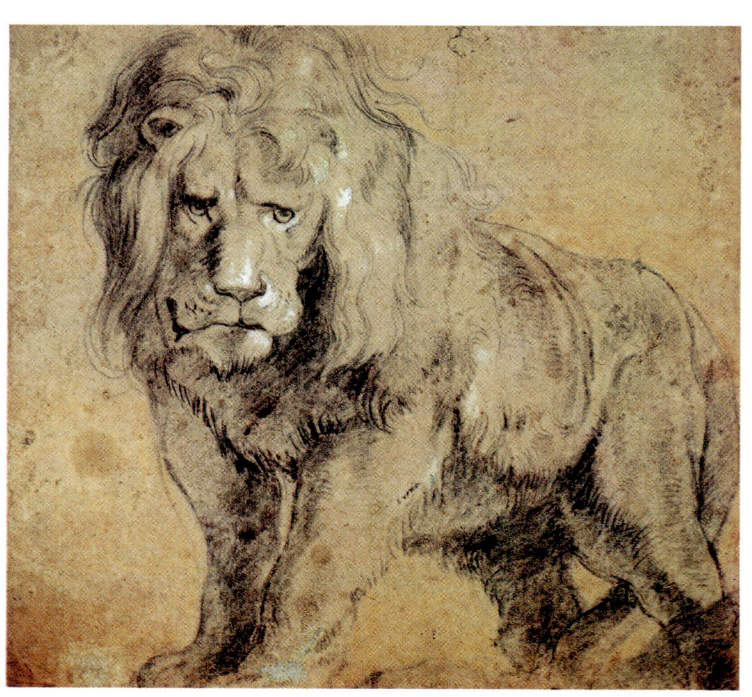

22. PETER PAUL RUBENS. *Lion.* c. 1614/15. Black chalk with white heightening and yellow chalk, 10 × 11 1/8″. Ailsa Mellon Bruce Fund

Peter Paul Rubens

One of the most renowned artists of his time, Peter Paul Rubens was born in Westphalia, a province of Germany, in 1577. When Rubens was ten, his family returned to their native Antwerp in what is now Belgium. He studied with several prominent painters in Antwerp and then spent eight years in Italy studying the great works of the Renaissance and classical antiquity. When he returned to Antwerp to marry, he was already known as a painter, scholar, diplomat, and master of seven languages. He was appointed court painter to the Governors of the Netherlands, a position he held for the rest of his life. Rubens received commissions and honors from all over Europe, and was even knighted by King Charles I of England. He died in Antwerp in 1640.

PETER PAUL RUBENS

Daniel in the Lions' Den

Lion

Peter Paul Rubens liked to paint large, imposing pictures in which he could allow his bold and vivacious brushwork its fullest freedom of expression. For the painting shown here, Rubens chose his subject from the biblical story of Daniel.

Daniel was chief counselor to Darius, king of Persia. The other counselors of the court were jealous of the young Hebrew and plotted against him. They convinced the king, much against his will, to have Daniel thrown to the lions. Darius was so disturbed by what he had done that he could not sleep or eat. At daybreak the morning after his order had been carried out, he hurried to the lions' den and had the stone rolled away from the entrance.

The amazed king found Daniel unharmed amid the lions, giving thanks to his God for his protection and deliverance. He commanded that Daniel be set free and that those who had accused him be cast into the lions' den in his place.

Darius was so impressed with Daniel's God that he sent out word to all his subjects, saying, "I make a decree, that in every dominion of my kingdom men tremble and fear before the God of Daniel: for he is the living God, and steadfast for ever."

Daniel is seated on his gracefully draped robe. His head is thrown back, his gaze turned upward toward the heavens, his hands folded in prayer. The lions are painted in muted golden-brown tones with delicate shadings of pink. Some of them squint and yawn in the early-morning sunlight flooding their lair, others snarl, and some appear perplexed by the miraculous event.

In the foreground, on the floor of the den, are a skull and the bones of previous victims. The dark area on the left underscores the somberness of the scene.

The artist drew several of the lions from life in the royal menagerie at Brussels; he took the lioness on the right from a Renaissance bronze statuette. For the figure of Daniel, Rubens borrowed from an Old Master drawing he owned. He skillfully combined and altered these borrowings to express the drama of the biblical story.

Rubens's drawing of a lion shown here may have been done in preparation for the artist's painting of Daniel in the lions' den. With its clean lines, the drawing evokes the full power of the beast. The head, with the curl and flow of the mane to either side of the staring eyes and sharply defined muzzle, is masterfully drawn with few lines and simple shading. Short strokes on chin, body, and legs are used to give the appearance of fur.

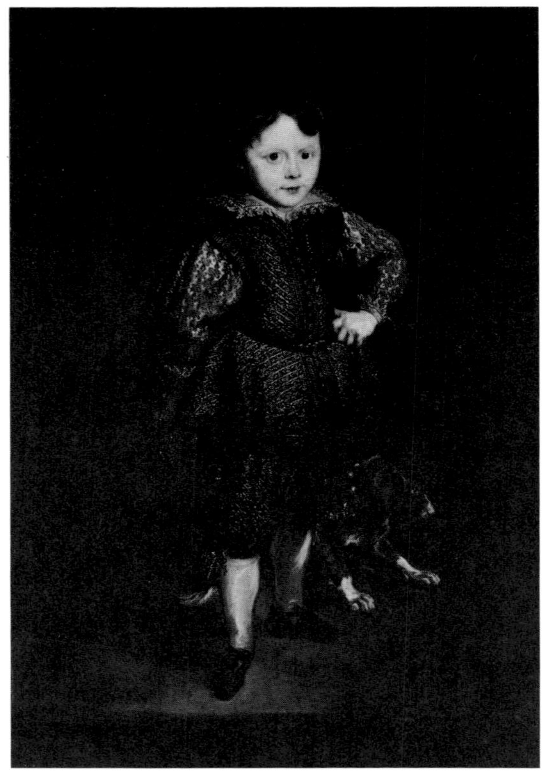

23. SIR ANTHONY VAN DYCK. *Filippo Cattaneo, Son of Marchesa Elena Grimaldi.* 1623. Canvas, 48 1/8 × 33 1/8″. Widener Collection

24. SIR ANTHONY VAN DYCK. *Clelia Cattaneo, Daughter of Marchesa Elena Grimaldi.* 1623. Canvas, 48 1/8 × 33 1/8″. Widener Collection

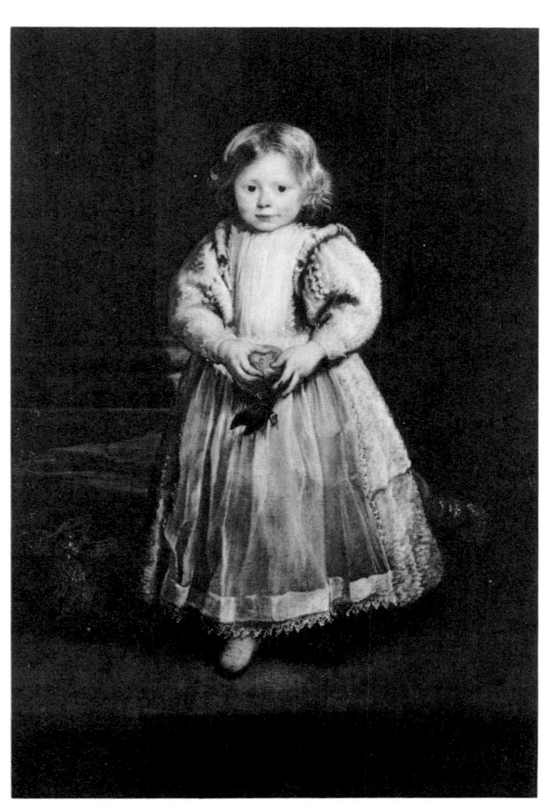

SIR ANTHONY VAN DYCK

Filippo Cattaneo, Son of Marchesa Elena Grimaldi

Clelia Cattaneo, Daughter of Marchesa Elena Grimaldi

Susanna Fourment and Her Daughter

Van Dyck has portrayed Filippo Cattaneo, at age four and a half, in the grand manner, wearing a fashionable Spanish cavalier's uniform. The son of Marchesa Elena Grimaldi and Marchese Nicola Cattaneo stands with his hand on his hip, one foot forward. The lavish gold-brocaded black costume with puffed brocaded sleeves in a lighter gold is relieved by a white lace-trimmed collar that sets off the child's face. Van Dyck has made us feel a certain impishness and jaunty vigor in the boy. Behind Filippo is a resigned-looking puppy; the boy grips his leash.

Although Filippo is set against a dark background, there is a bright golden glow to the picture.

Clelia Cattaneo, Filippo's sister, is pictured standing at the top of a step in front of an upholstered stool. In the background, the shadowy outlines of a column can be seen. There is a trace of self-consciousness and apprehension in the child's face. She clutches the fruit in a gesture that is wholly natural and childlike. Her figure is bathed in light.

Susanna Fourment came from a middle-class family of Antwerp. Her father was a silk merchant.

In Van Dyck's portrait she is seated in an armchair on a columned terrace. Her small daughter stands at her side. Large folds of red drapery in the background accent the white, lace-trimmed double ruff Susanna wears about her throat. Her brownish-red hair, which is combed off her high, wide forehead, is partly covered by a jeweled cap. Expressive brown eyes and delicate features lend distinction to her small oval face. The daughter, who clasps the mother's large hand with her small ones, is, in her dress of shimmering orange-red silk, as elegantly clothed as her mother. She bears a strong resemblance to Susanna in her delicate features as well as in her alert expression. In his portraits,

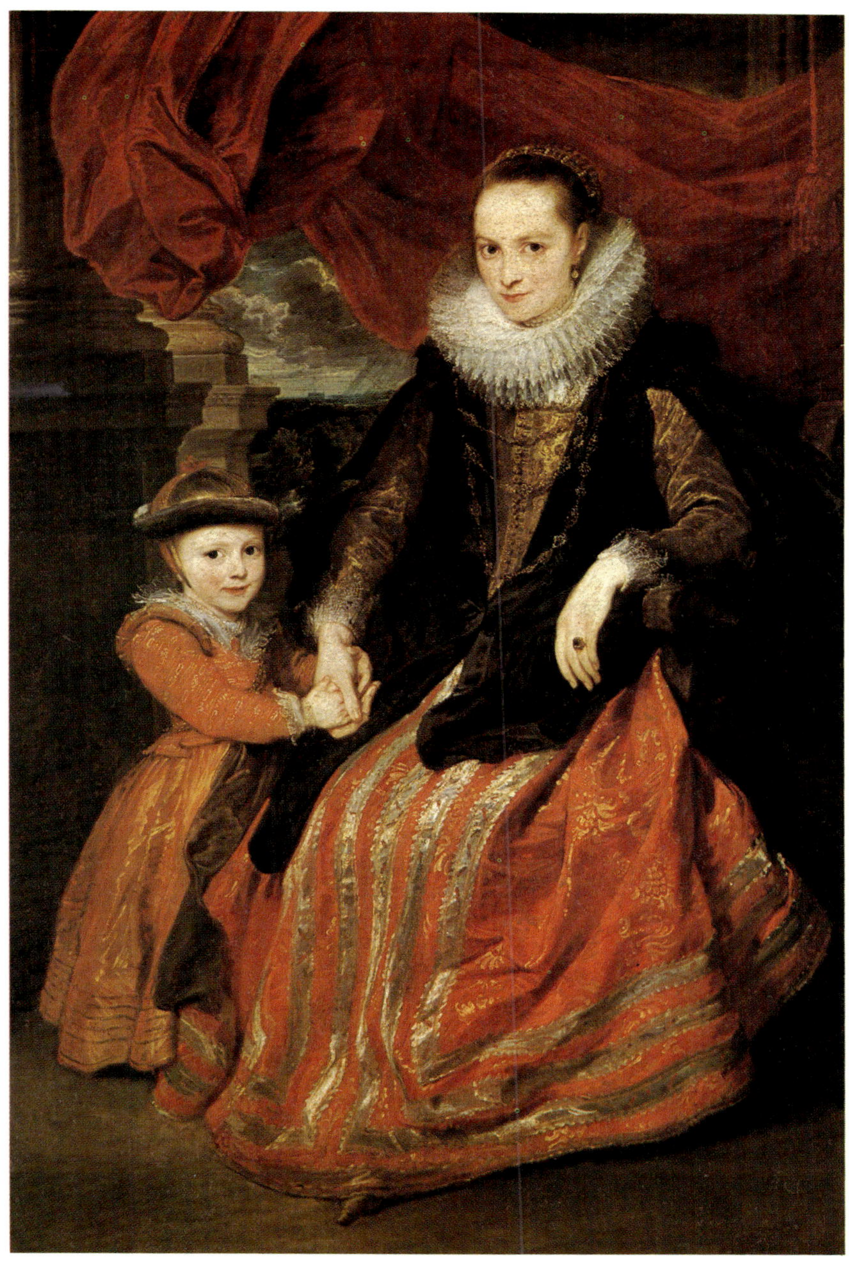

25. SIR ANTHONY VAN DYCK. *Susanna Fourment and Her Daughter.* c. 1620. Canvas, 68 × 46 1/4″. Andrew W. Mellon Collection

Sir Anthony van Dyck

Anthony van Dyck, who was born in Antwerp in 1599, was apprenticed to a painter at the age of ten. By the time he was sixteen, he had his own studio, and soon afterward he became a full member of the Guild of St. Luke, an unusual honor for so young an artist. After working as an assistant to Peter Paul Rubens, Van Dyck spent six years studying the old masters in Italy, then settled down to paint in Antwerp. In 1632 he was invited to England by King Charles I, who later knighted him "Sir Anthony Vandike, principalle Paynter in ordinary to their Majesties." Sir Anthony van Dyck painted portraits of the royal family and other members of the nobility, and died in England in 1641.

Van Dyck was more concerned with creating a feeling of decorative elegance than with revealing the feelings and personalities of his sitters. The refinement of the painter's style, his love of lavish materials, and his fluid and smooth brushstrokes lend stateliness and grace to all his subjects.

The artist did these three portraits in the early 1620s in Genoa, Italy. They have remained together for over three hundred years. Until 1906 they enhanced the walls of the richly decorated Cattaneo Palace in Genoa. Shortly thereafter they became a part of the private collection of Mr. P.A.B. Widener in Philadelphia, Pennsylvania. Since 1942 they have hung as a group in the Widener Collection in the National Gallery.

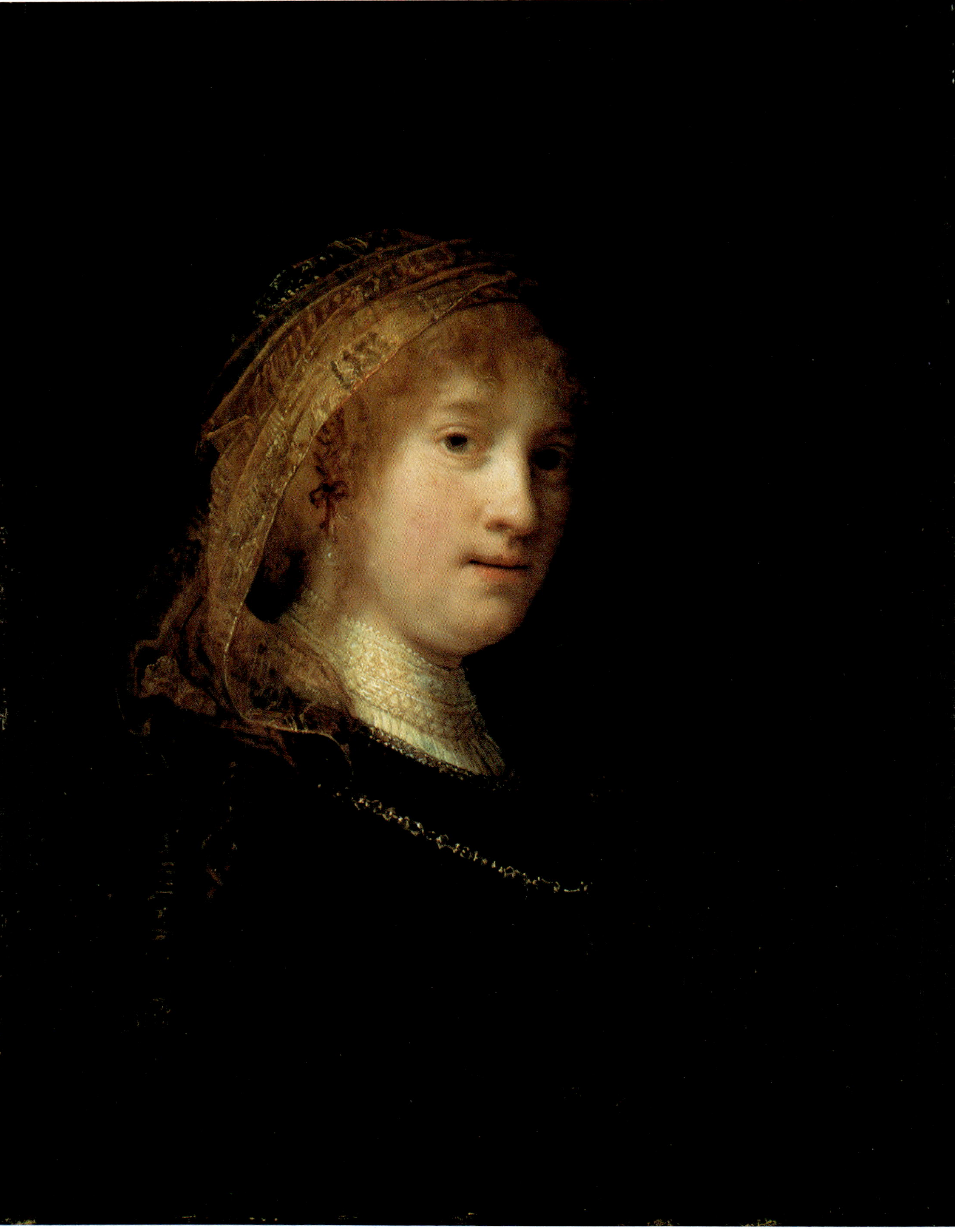

26. REMBRANDT VAN RIJN. *Saskia van Uilenburgh, the Wife of the Artist.* c. 1633. Wood, 23 3/4 × 19 1/4″. Widener Collection

REMBRANDT VAN RIJN

Saskia van Uilenburgh, the Wife of the Artist

In this painting done shortly before their wedding, Rembrandt has portrayed Saskia, his wife-to-be, in the bloom of her youth, looking content and confident.

Saskia is in profile with her head turned slightly toward the viewer. A sheer gold-embroidered veil covers her auburn hair and falls over her shoulders. She is attired in a dark blue dress with a richly embroidered and finely pleated white inset neckpiece edged with narrow gold trimming. She wears a long heavy gold chain and a gold earring with a large pearl drop. Rembrandt has placed her against a dark background; exquisite light floods her attractive face, which has the suggestion of a smile.

Saskia died seven years after her marriage to Rembrandt. After her death the artist began to concentrate more on conveying the inner thoughts and feelings of his sitters and less on their outward appearance, often to their displeasure and to his financial loss.

Rembrandt van Rijn

The son of a miller, the great Dutch artist Rembrandt van Rijn was born in 1606 in Leiden, the Netherlands. Rembrandt began painting at the age of fifteen, and quickly displayed his genius in many small paintings and etchings of historical and religious subjects. Rembrandt moved to Amsterdam in 1631, where his group portrait of surgeons, *The Anatomy Lesson of Dr. Nicolaas Tulp*, soon assured his success. His wife, Saskia, was wealthy, and he received many commissions, but his fortunes declined after Saskia's death, and he died in poverty in 1669. Many of his finest and most moving works belong to these last difficult years.

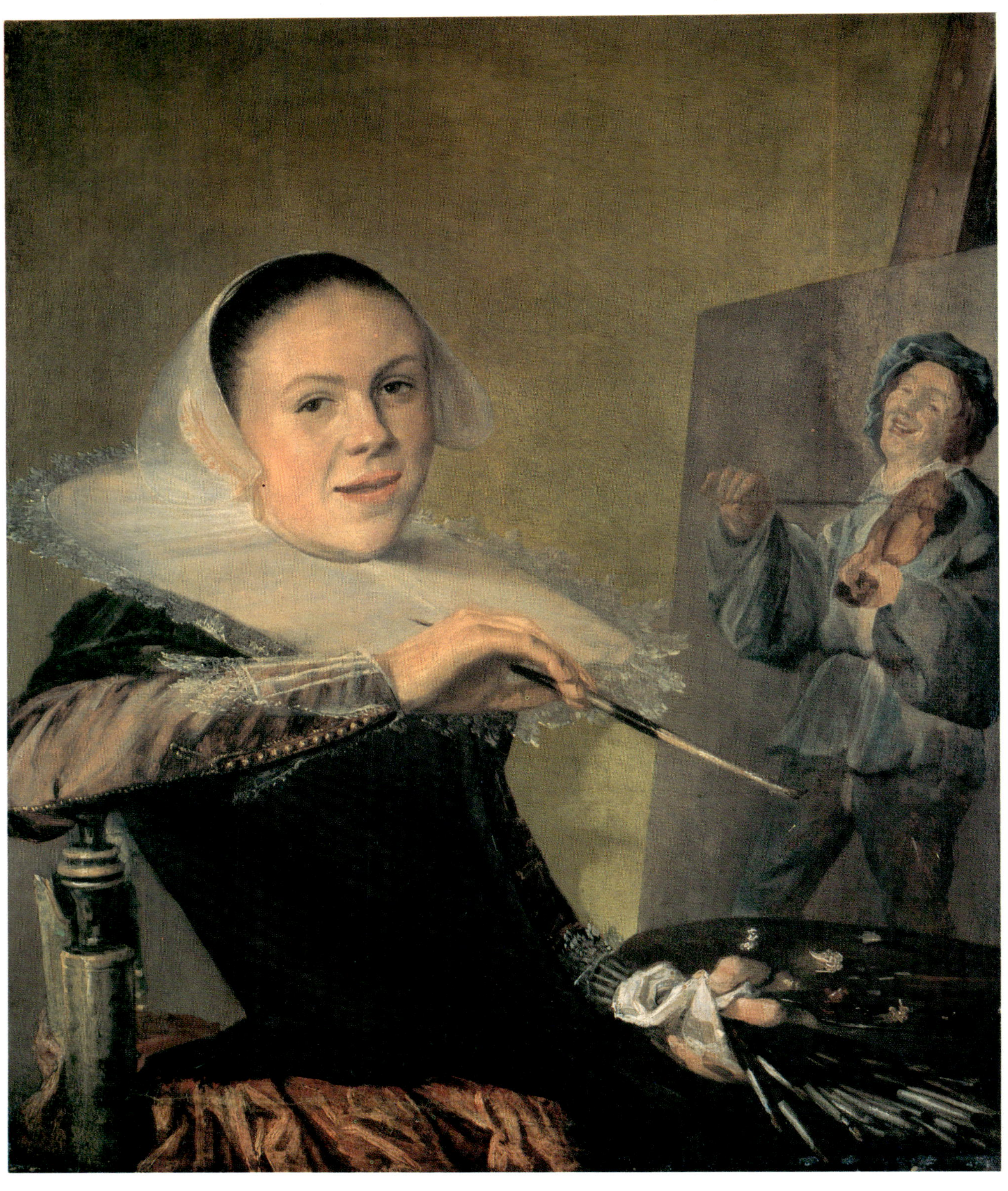

27. JUDITH LEYSTER. *Self-Portrait.* c. 1635. Canvas, 29 3/8 × 25 5/8″. Gift of Mr. and Mrs. Robert Woods Bliss

Judith Leyster

One of the first women to gain recognition as a painter, Judith Leyster was born in the Dutch city of Haarlem in 1609. She became a pupil of Franz Hals, a leading Dutch painter, and eventually established her own workshop. Leyster specialized in dramatically lighted, intimate pictures using few figures. She signed all her works with the initials "JL" and a star, since the Dutch word *leysterre* means "lodestar." She married Jan Molenaer, a genre painter, and continued to paint successfully until her death in 1660.

JUDITH LEYSTER

Self-Portrait

JEAN-BAPTISTE-CAMILLE COROT

The Artist's Studio

Both the seventeenth-century Judith Leyster *Self-Portrait* and the nineteenth-century painting by Corot shown here portray artists' studios and both depict paintings within the paintings. But whereas Leyster concentrated on her self-portrait, Corot was more interested in the objects in the studio and the patterns they created.

In Judith Leyster's painting the artist sits facing the viewer. On the right is an easel showing the almost completed portrait of a laughing young man in blue playing a violin. The artist's right arm is balanced on the backrest of the chair; she gracefully holds a poised paintbrush. In her left hand she clasps her palette, with its dabs of colorful paint, and a wiping cloth.

The different shades of beige and tan used in the picture create a feeling of richness and warmth. Leyster portrays herself with a pleasant but not beautiful face, parted lips, and large dark amused eyes. She looks out at us, as if beckoning us into the painting. We feel that here is a person at ease with herself, her chosen profession, and the world.

The Corot painting is rendered in soft, muted shades with a few strong, dark outlines. Seated on a wooden chair, turned to face the easel, is a young woman. On the easel is a landscape by Corot which she adjusts with her hand while she carefully studies the work. In her other hand she holds a mandolin, resting it on the floor. The dog approaching her chair is rendered without strong outlines; its dark form is blurred. In the background the black pipe from a coalstove runs up the wall, forming a strong vertical accent in the painting. Dimly seen on the floor to the right against the wainscoting is a large painting. Small landscapes, statuettes on wall brackets, and a figure painting are hung on the wall above the wainscoting. Our eye takes in the scene quickly, for the artist has placed all forms in a clear geometric arrangement and has painted them solidly. Yet there is a feeling of gentleness in the picture which stems from the curving form of the young woman, set off against the verticals and horizontals of the room, and from the soft silvery light which bathes the forms and unifies them.

In this work Corot seems to be contrasting the paintings for which he won public acclaim—those displayed on the wall—with those he did for his own pleasure—represented by the landscape on the easel. Today his fame rests in great part on the landscape paintings he so loved.

The picture shown here must have been very precious to the artist. Until his death it hung in his living room. Another, later example of his beautiful figure compositions is seen in *Gypsy Girl with Mandolin* (plate 19).

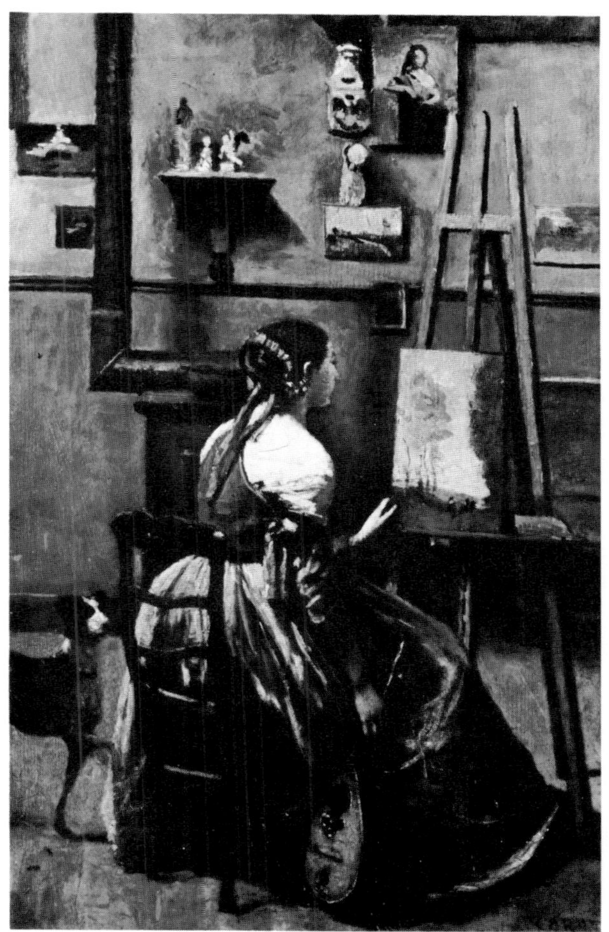

28. JEAN-BAPTISTE-CAMILLE COROT. *The Artist's Studio.* c. 1855/60. Wood, 24 × 15 1/2". Widener Collection

SEE PAGE 41 FOR COROT'S BIOGRAPHY

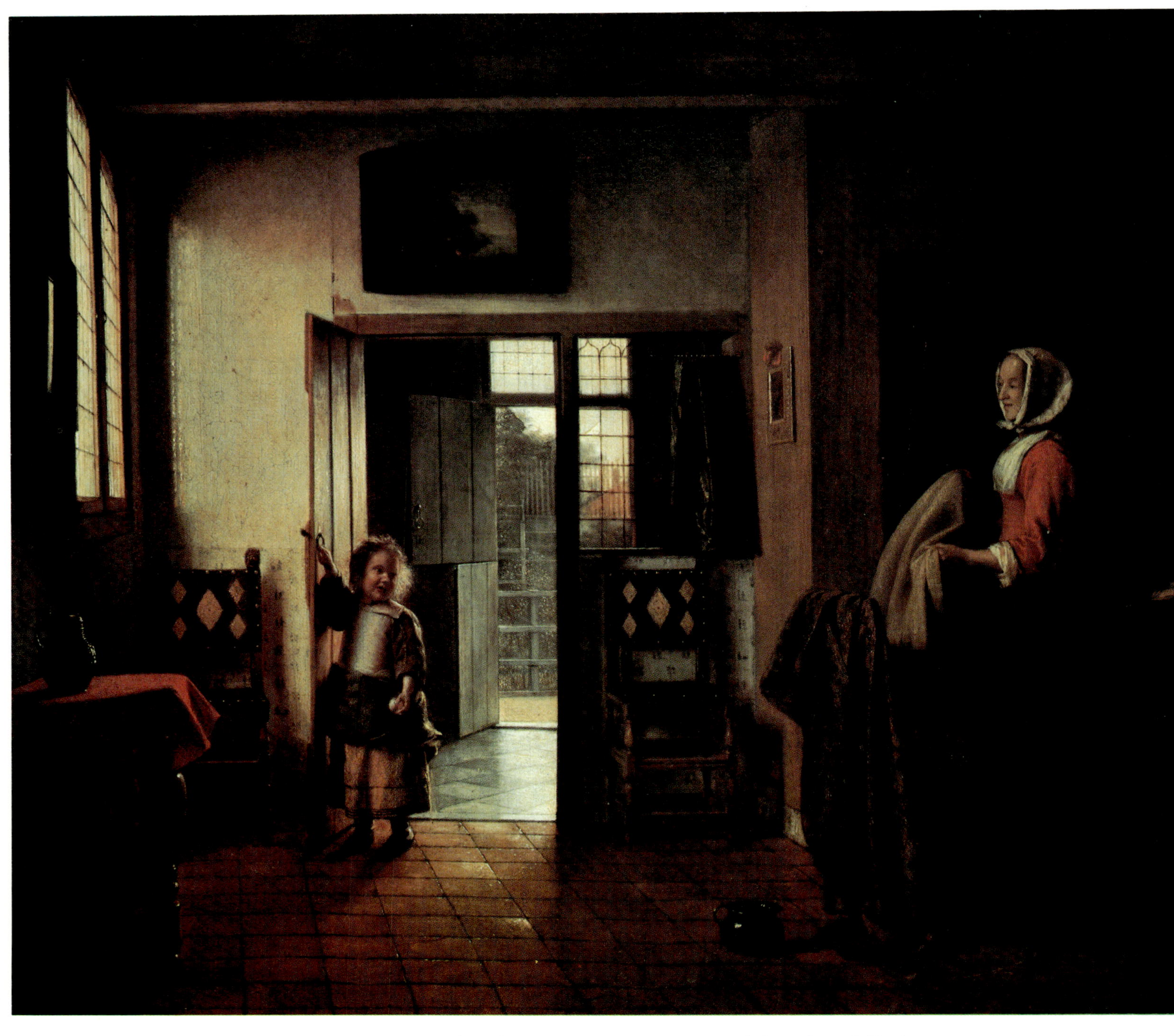

29. PIETER DE HOOCH. *The Bedroom.* c. 1660. Canvas, 20 × 23 1/2″. Widener Collection

Pieter de Hooch

Though Pieter de Hooch is much admired as an artist, little is known about his life. The son of a mason, he was born in Rotterdam, the Netherlands, in 1629. When he settled in the city of Delft as a young man of twenty-six, he was already an accomplished painter of middle-class Dutch interiors. It was in Delft, where he married and had two children, that De Hooch did most of his finest work. After the death of his wife, De Hooch moved to Amsterdam, where he died about 1684.

PIETER DE HOOCH

The Bedroom

JAN VERMEER

A Lady Writing

Certain Dutch painters of the seventeenth century are called "Little Masters." They were superlative craftsmen and are called "little" only because they limited themselves in the type of subjects they pictured. They were primarily painters of landscapes, portraits, interior scenes, and also of "still lifes," inanimate objects such as flowers, fruit, and shells. As a group their pictures reflect the daily life of a people as the art of no other period or area has done.

Among the best of the Little Masters is Pieter de Hooch. In *The Bedroom* this artist has, as usual, shown a very simple event: a child interrupts her mother, who is making the bed in the curtained alcove behind her. De Hooch's wife and daughter are undoubtedly the two figures portrayed. They appear often in his paintings, as does the room itself, which surely was in the artist's home. The picture tells us much about the home, costume, and manner of life of the middle-class Dutch of the time; but the artist's interest was not as much in the scene itself as in creating a feeling of light and air in his paintings. In this De Hooch excelled. The interior is so realistically rendered that we feel that we are in the room with the mother and child.

Jan Vermeer, who painted *A Lady Writing*, was similar in many respects to Pieter de Hooch. Both painters were Dutch, were a few years apart in age, lived and painted in Delft, appear to have used members of their families as models, and were recognized as accomplished artists during their lifetimes. Very little about the life of either artist is known today.

A Lady Writing shows a richly dressed young woman. Her jacket is lavishly trimmed with ermine fur. The leather-backed chair on which she sits has two carved lion heads. The writing table is covered with a rich velvet cloth upon which are some writing materials, a strand of pearls, and a small decorated chest. The woman, who holds a quill pen in her hand, leans forward and looks out as if she is thinking about what to say next to her correspondent.

The painting is very realistic and conveys a sense of quiet peacefulness. One can almost feel the softness of the flesh and the textures of the various materials. Vermeer carefully studied how light illuminated figures and objects and was reflected by them. His sensitive rendering of light effects combined with meticulous craftsmanship gives particular distinction to his work.

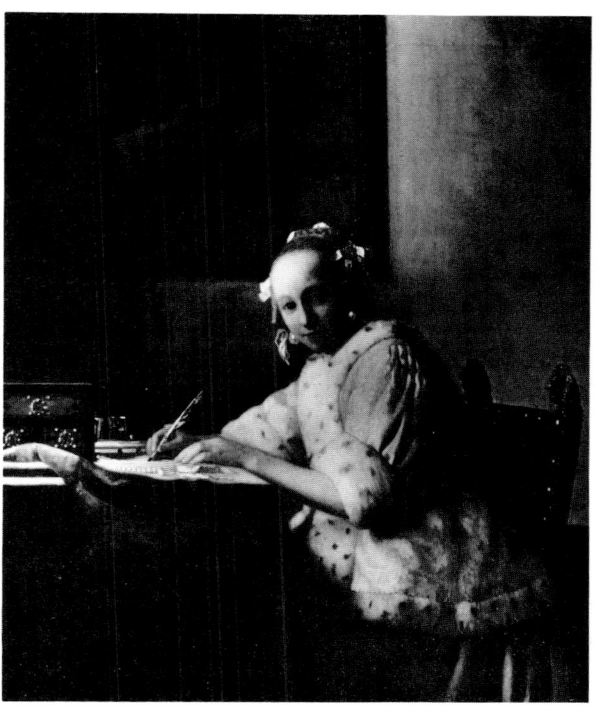

30. JAN VERMEER. *A Lady Writing*. c. 1665. Canvas, 17 3/4 × 15 3/4". Gift of Harry Waldron Havemeyer and Horace Havemeyer, Jr., in memory of their father, Horace Havemeyer

Jan Vermeer

Jan Vermeer was born in 1632 in the thriving Dutch city of Delft, where he apparently spent his entire life. He became a member of the Painters' Guild and devoted his life to his art and family, which is believed to have been very large. A painstaking craftsman, Vermeer worked slowly, producing fewer than forty paintings in his lifetime. His favorite subject was softly lit Dutch interiors. Though his work had a ready market and commanded substantial prices, Vermeer was plagued with financial problems until his death in 1675.

31. GILBERT STUART. *The Skater*. 1782. Canvas, 96 5/8 × 58 1/8″. Andrew W. Mellon Collection

GILBERT STUART

The Skater

HENDRICK AVERCAMP

A Scene on Ice

Gilbert Stuart's *The Skater,* or "A Gentleman Skating in St. James' Park," as it was first called, is both noble and powerful. The skater, a Scot named William Grant, is dressed in black from head to toe. This somber costume is relieved only by a white stock tie, the gray fur lining the lapels of his coat, and the silver buckles on his skates. His handsome manly face has a ruddy color which, along with the bleak wintry sky and the bare trees, gives us a sense of the coldness of the day. His arms are crossed, accenting the power of his broad shoulders. The small skating figures in the middle distance of the painting, those standing or seated on the bank to the right, and the hazy towers of London's Westminster Abbey in the background contrast with the clear figure of the skater and make him appear larger than life-size.

When the painting was first exhibited at the Royal Academy in London in 1782, it brought instant fame to the twenty-seven-year-old Stuart, and to Grant, who went to the exhibition dressed in his skating costume and was followed everywhere by enthusiastic crowds.

In *A Scene on Ice* the Dutch artist Hendrick Avercamp shows a typical winter scene in Holland over three hundred years ago.

The artist has dressed some of his tiny figures in dark clothes, others in bright reds, soft greens, light browns. The reds—particularly in the skirts of the two women in the foreground—stand out against the gray of the ice and the wintry sky.

On the right two boys play at an informal game of ice hockey. People wearing skates with long runners skim over the ice, while others walk across the frozen waters. In the center an elegant plumed horse guided by a groom pulls a hand-painted sleigh carrying two wealthy ladies.

Two dogs frolic in the foreground. One, beside the man who is refastening his skates, eyes the other running toward him. Near the first dog, as part of the design, the artist has signed his painting with his initials, "H.A."

Farther back are many other figures. Some are going about their daily tasks: pushing sleds filled with wares they have bought or are planning to sell.

The sky, painted in the same tones as the ice, is dotted with a few birds. To build up his scene Avercamp has added the soft brownish-gray forms of sturdy dykes, snow-covered houses, buildings, and a windmill.

What could be more delightful, gay, and interesting than this colorful winter landscape!

Gilbert Stuart

Gilbert Stuart was born in Rhode Island in 1755. At fourteen he was painting portraits, and his work so impressed a visiting Scottish artist that he was taken to Edinburgh briefly to study. Later, he became a success in London and Ireland, but was so extravagant with his money that he was forced to flee both countries to avoid debtors' prison. Back in America, he painted several portraits of President George Washington; these are among the best-known images in the history of American painting. Stuart finally settled in Boston, where he died in 1828.

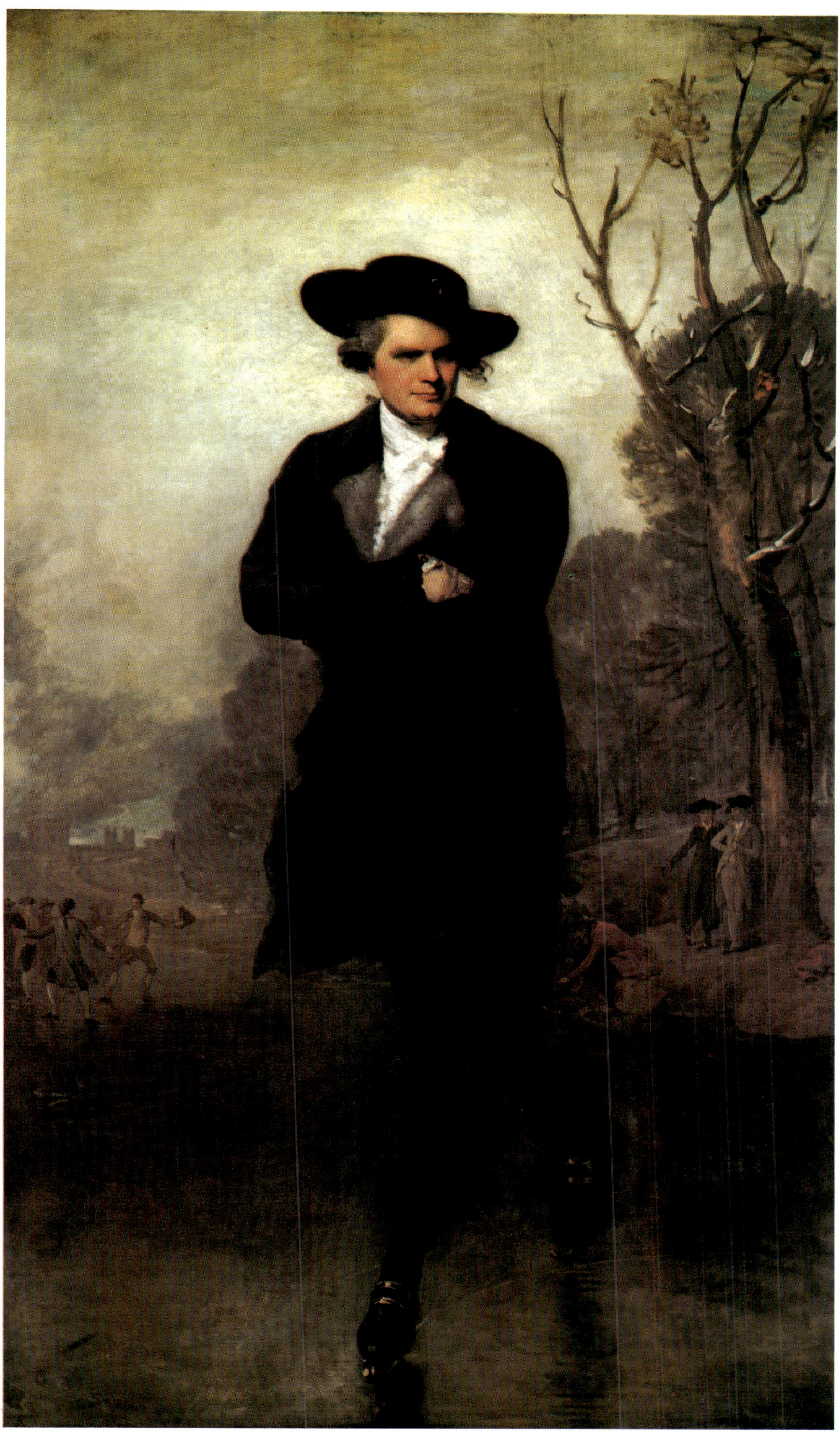

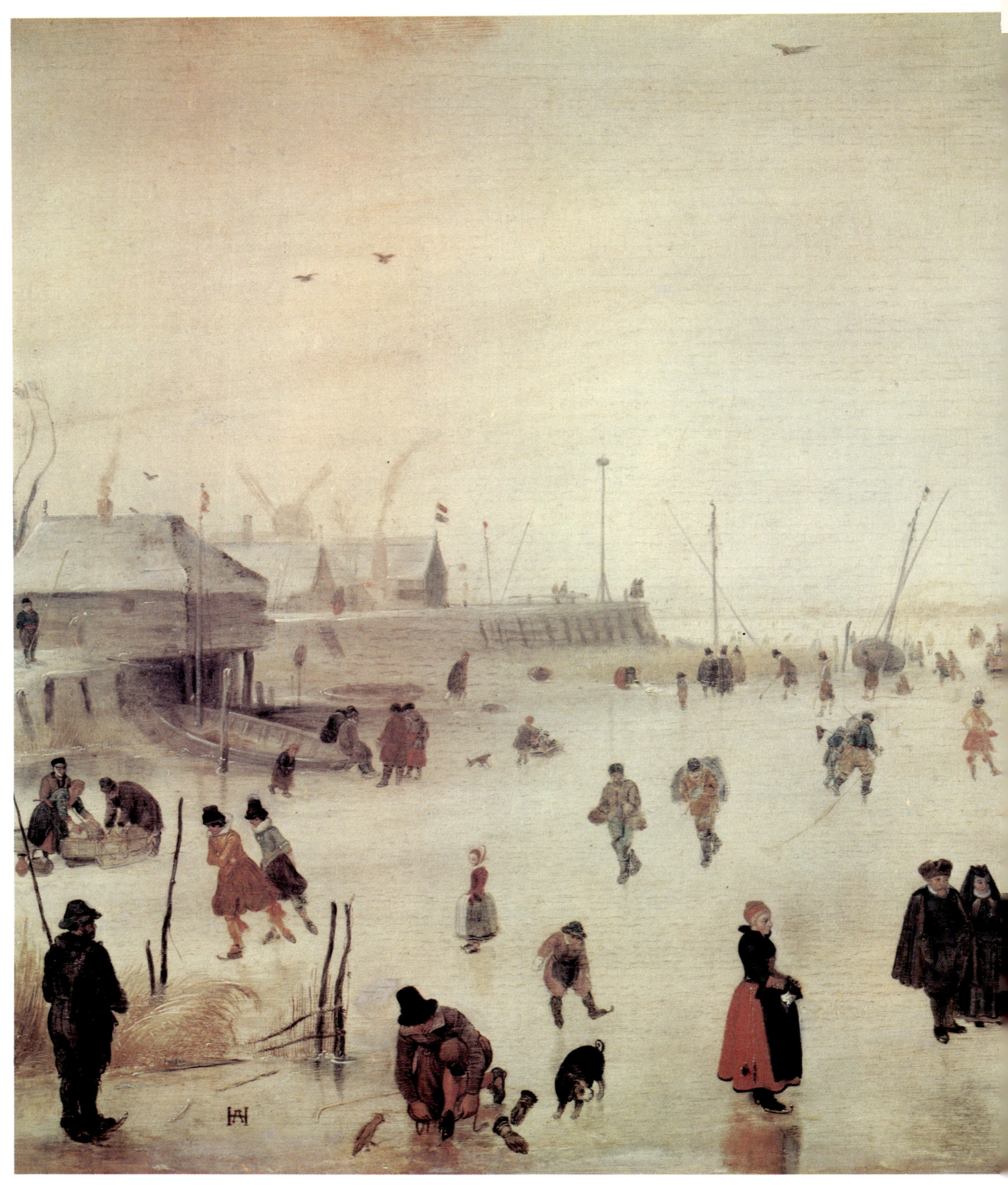

32. HENDRICK AVERCAMP. *A Scene on Ice*. c. 1625. Wood, 15 1/2 × 30 3/8″. Ailsa Mellon Bruce Fund

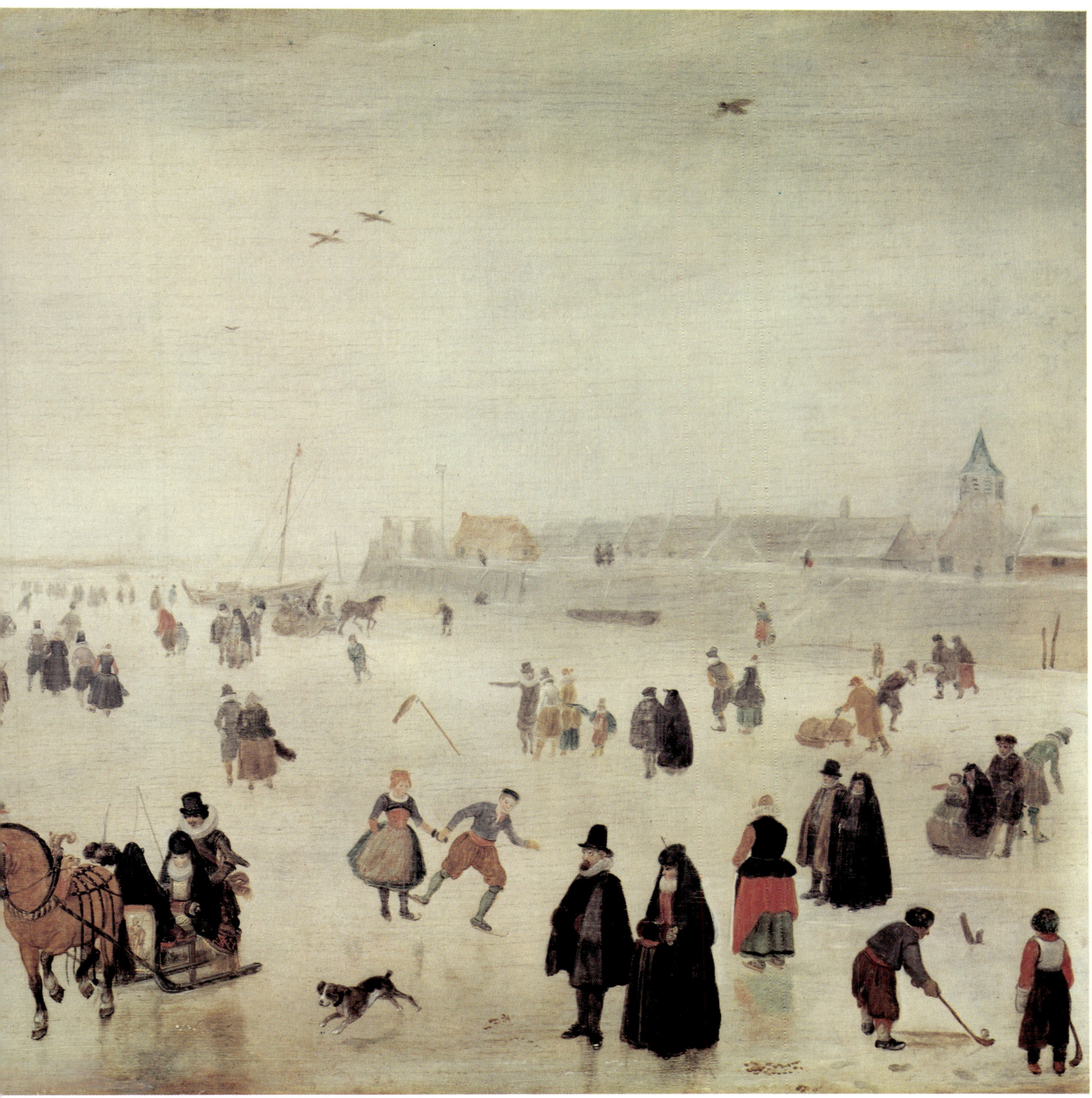

Hendrick Avercamp

Hendrick Avercamp was born a deaf-mute in Amsterdam in 1585. He settled down to paint in the quiet town of Kampen on the eastern bank of the Zuider Zee. Because of his handicap, he became known as "The Mute of Kampen." Avercamp was probably the first artist to specialize in winter scenes of outdoor sports. His paintings give us an idea of what Dutch life was like in the seventeenth century. Avercamp died in Kampen in 1634.

33. JAN STEEN. *The Dancing Couple*. 1663. Canvas, 40 3/8 × 56 1/8″. Widener Collection

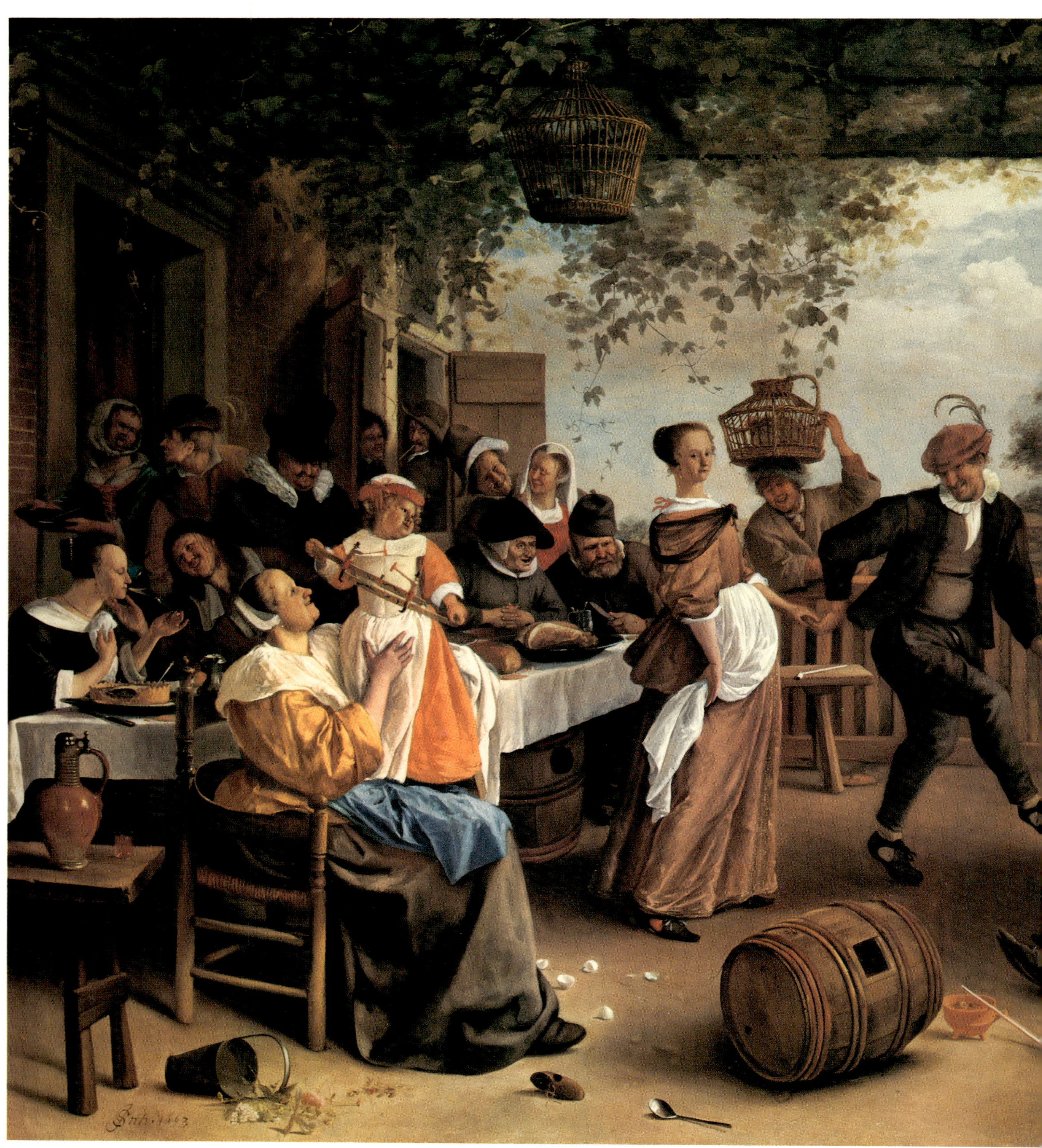

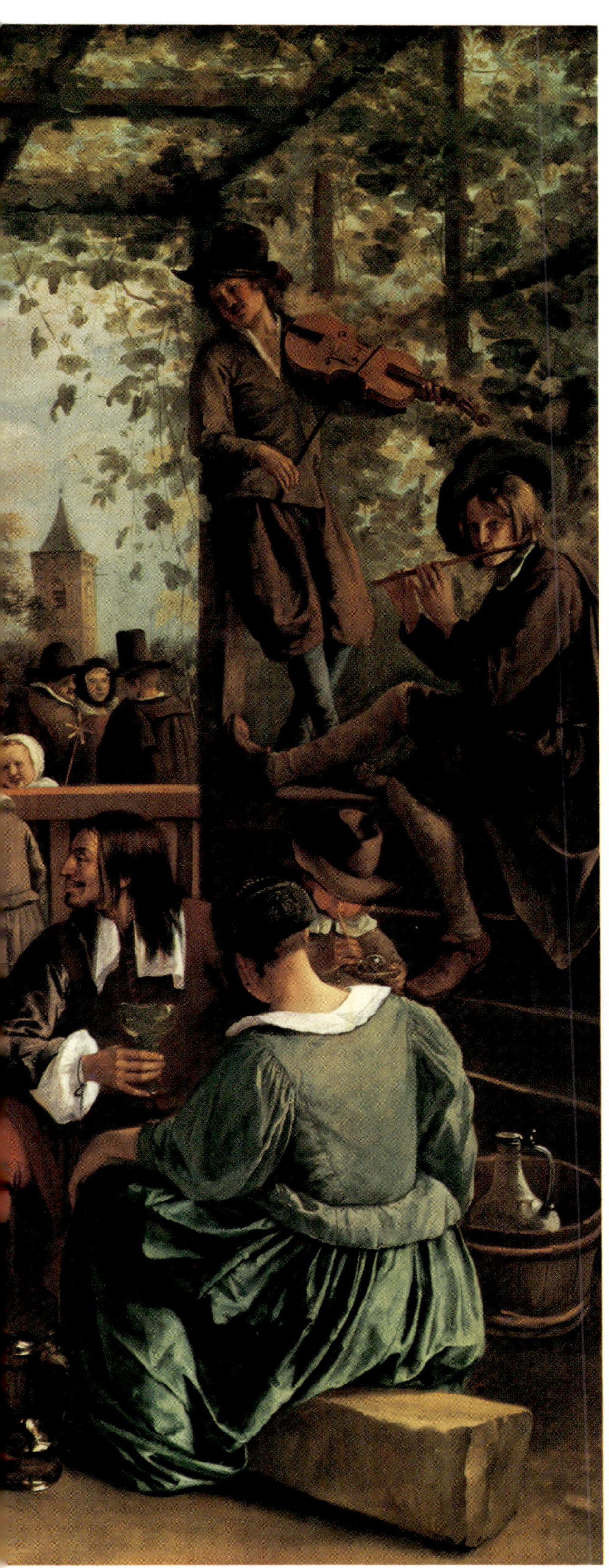

JAN STEEN

The Dancing Couple

Jan Steen's *The Dancing Couple*, a rollicking scene under the arbor of a tavern courtyard, is a fine example of the genre painting (depictions of scenes of daily life) so popular in Holland in the seventeenth century.

In the center of the picture, a timid young woman stands hesitating while the young man holding her hand has already started to move to the spritely music. Two young musicians with violin and flute are poised on the rail and steps at the right. Two couples sit at a table laden with food. The man affectionately patting the chin of his companion is thought to be Steen himself. The other couple are elderly. Behind them are a young man and maid, several other guests, and servants, all involved in the general festivity. Nearer to the viewer a woman balances a child in her lap; the brightly clothed youngster holds a wooden toy, but seems more interested in looking toward the couple seated on the log at the right—or perhaps he is watching the young child, his face partially covered by a large hat, who is blowing soap bubbles. The ground of the courtyard is strewn with broken eggshells, an overturned bucket of flowers, a keg, a spoon, a shoe, a pipe, and a pewter flask. Behind the fence a laughing man with a poultry basket on his head watches the fun. Booths of a village fair and another group of people can be made out in the background near the church.

In this gay, crowded, and colorful scene, Jan Steen has skillfully, humorously, yet compassionately caught some of the spirit of middle-class Dutch life of his time.

Jan Steen

Like Rembrandt, Jan Steen was born in Leiden, the Netherlands, but was younger than that famous artist, being born about 1626. Steen's father ran a brewery, and the artist also managed a brewery and was an innkeeper at different times in his life, though he maintained an active career as a painter of genre scenes. Steen lived in several Dutch towns with his family. After his wife died, he returned to Leiden and eventually remarried. One of the founders of the local Painters' Guild, Steen eventually became its dean. He pursued his many interests with unbounded vigor until his death in 1679.

CANALETTO

Venice, the Quay of the Piazzetta

Unlike most old European cities, Venice was not surrounded by fortified walls. Its lagoon provided both a safe harbor and protection against enemies.

Here Canaletto, the great Venetian painter of cityscapes, has given us a view of the main entrance to the city, the beginning of the Grand Canal and the quay (wharf) of the *piazzetta*, or little piazza, as it looked in the eighteenth century. Across the Grand Canal, silhouetted against the bright blue sky, is the double-domed church of Santa Maria della Salute. In the distance against the skyline is the Church of the Redeemer. The pale yellow stones of the walls of the harbor, the balustrades and buildings, the terra-cotta tile roofs, and the domes of the churches gleam in the sunlight.

In the lagoon is a three-masted merchant vessel flying the Union Jack, the British flag. Gondolas, the "water-taxis" of Venice, glide back and forth in the canal. The white, strangely shaped prows of other gondolas line the quay, waiting to ferry people to other parts of the city. The fishmongers near the center of the painting are selling their fresh catch to customers. Scattered here and there are tourists, mainly wealthy gentry from England taking the "Grand Tour" of Europe, a fashionable practice during the eighteenth century. The picture conveys a sense of ordered, leisurely activity.

Canaletto's scenes are remarkable for their accuracy and attention to detail. To create such painstakingly accurate paintings, Canaletto recorded his impressions in detailed pencil or pen-and-ink sketches made at the scene. He then took these sketches to his studio, where he produced exact, finished designs. Then he transferred these designs to canvas. The artist began by painting all the buildings and scenery. After completing the background, he added people and animals to fill out the picture.

So precise is the painting that the viewer can even tell the time of day. From the fact that the long shadows cut directly across the picture at the front from the southwest, we can determine that the sun is in the southeast; thus the time is about an hour before noon.

On the landing stage of the quay is a stone engraved with the letters *A. C. F.*, meaning Antonio Canaletto *Fecit* or "made" this picture. In this way Canaletto put his signature on the painting but disguised it as part of the wall. The picture is so convincingly real that we feel we might step through it into the past and become part of the life of this fascinating city.

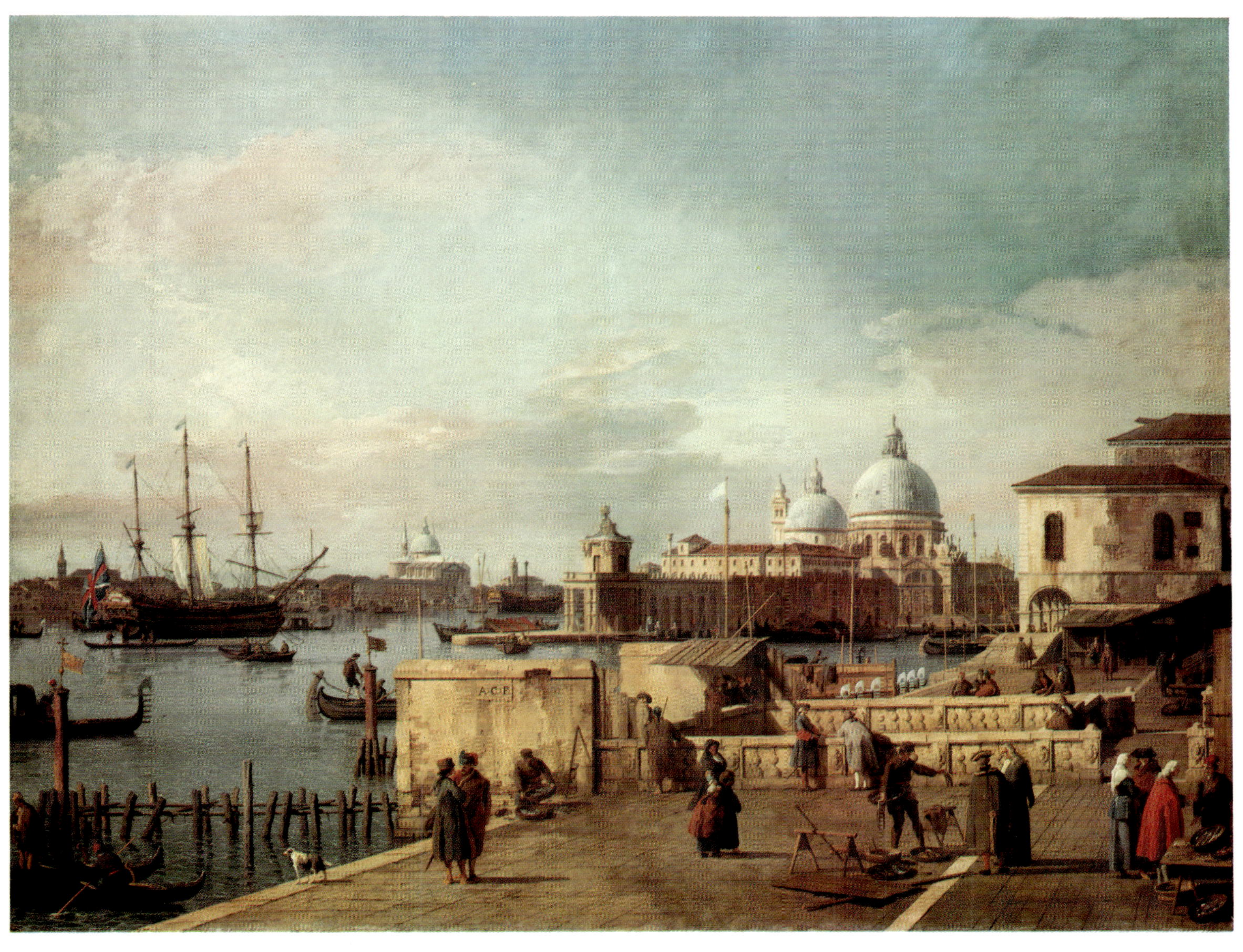

34. CANALETTO. *Venice, the Quay of the Piazzetta.* Early 1730s. Canvas, 45 1/8 × 60 3/8″. Gift of Mrs. Barbara Hutton

Canaletto

Giovanni Antonio Canal, known as Canaletto, was born in 1697 in Venice. His father painted stage sets for operas, and when Antonio was old enough he became his assistant. Antonio soon began to paint views of Venice to sell to foreign tourists, changing his name to Canaletto, "Little or Younger Canal," to avoid confusion between himself and his father. Canaletto's pictures became so popular in England that he was invited there in 1745 to paint Britain's landmarks. He remained almost without interruption until 1755. Canaletto was elected to the Venice Academy and maintained an active studio until his death in 1768.

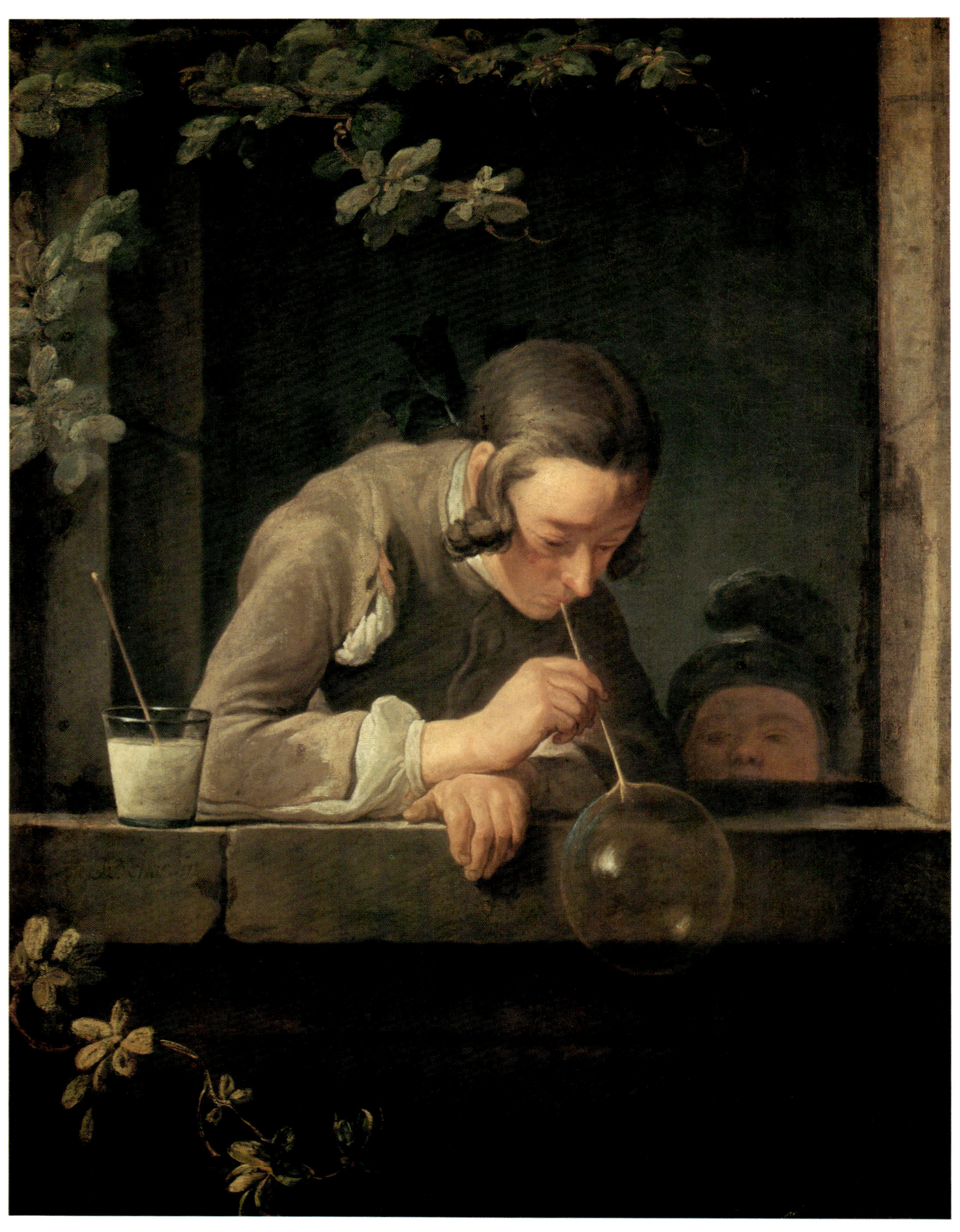

35. JEAN-BAPTISTE-SIMÉON CHARDIN. *Soap Bubbles*. c. 1745. Canvas, 36 5/8 × 29 3/8″. Gift of Mrs. John W. Simpson

JEAN-BAPTISTE-SIMÉON CHARDIN

Soap Bubbles

The House of Cards

The French painter Chardin was a great admirer of Dutch painters such as De Hooch and Vermeer (see plates 29, 30). Like these painters he chose his subjects from everyday middle-class life, and painted them with a simplicity and clarity of design which accent their solid reality.

A boy leaning out of a window and blowing bubbles may have seemed unimportant to many painters of Chardin's day. They were primarily interested in representing the pastimes of ladies and gentlemen. Yet notice what exacting care the artist has taken in depicting this scene, probably one he had witnessed many times. He has devoted himself to a careful and sympathetic study of textures and tones and harmoniously related colors. He has concentrated on a small area so intently that we feel the exclusion of the outside world. We could remain forever with the little boy, who strains over the windowsill to watch the luminous bubble that will never burst.

In *The House of Cards* Chardin has conveyed the same feeling of total absorption we sense in *Soap Bubbles*. The young boy, his long hair tied with a ribbon in the fashion of the time, is perhaps a servant who has been sent into the game room to tidy up. Intrigued by the folded cards, the lad has seated himself at the gaming table, upon which are scattered several chips, and has begun to build his house. Chardin portrays the boy's activity in a serious and dignified manner. We can feel his intense concentration; he seems to be unaware of the open drawer with the protruding Knave of Hearts plainly in view.

The boy's face, the cards, and the edges of the table are highlighted. Chardin has placed his subject against a neutral background. The overall effect is one of calmness and order.

It was in the fourteenth century that playing cards found their way to Europe from the East. Many people believed that they were an invention of the devil to cause people to waste time. Nevertheless, by the eighteenth century card-playing had become a popular pastime. The backs of the cards were left plain lest they be marked by cheaters. Why are the cards creased? It was the custom at the time to bend them after a game was finished, to avoid trickery in their reuse. The Knave of Hearts symbolized rascality and may reflect the artist's negative view of card-playing.

In both of these works, the viewer feels a kinship in having participated in these same simple pleasures.

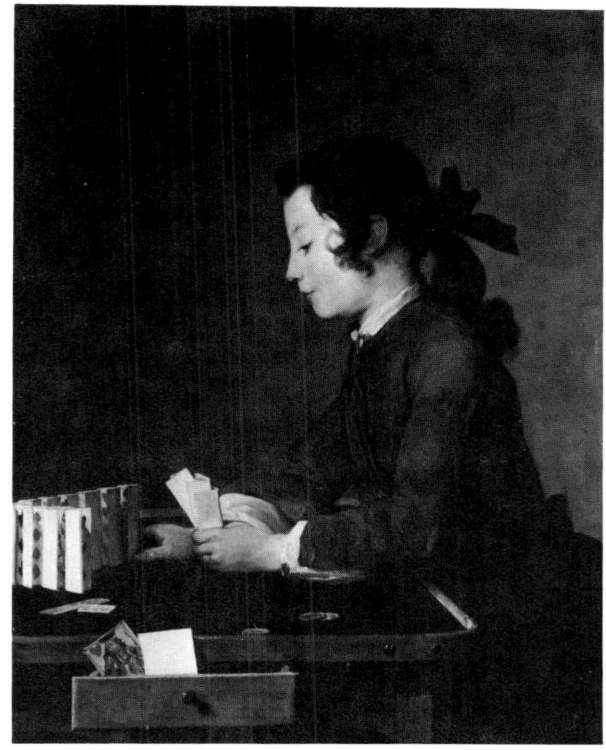

36. JEAN-BAPTISTE-SIMÉON CHARDIN. *The House of Cards.* c. 1735. Canvas, 32 3/8 × 26″. Andrew W. Mellon Collection

Jean-Baptiste-Siméon Chardin

The son of a cabinetmaker, Jean-Baptiste-Siméon Chardin was born in Paris in 1699. He worked as an apprentice in his father's shop, but decided very early to become a painter. Chardin most enjoyed painting still lifes, but it was his pictures of the daily life of the typical middle-class Parisian household that established his reputation. While still a young man he was made a member of the Royal Academy, and later Louis XV honored him with royal patronage. Chardin died in 1779.

37. JEAN-HONORÉ FRAGONARD. *A Young Girl Reading*. c. 1776. Canvas, 32 × 25 1/2″. Gift of Mrs. Mellon Bruce in memory of her father, Andrew Mellon

JEAN-HONORÉ FRAGONARD

A Young Girl Reading

When Jean-Honoré Fragonard painted the charming portrait of a young girl shown here, a great change had taken place in French art. Artists now brought more light, color, delicacy, rhythm, softness, and liveliness to their work.

In his earlier years as a painter, Fragonard was best known for his decorative scenes, pictures of people playing popular eighteenth-century games and of noblemen and ladies enjoying themselves in rainbow-colored landscapes with graceful trees, beautiful gardens, fountains playing under billowy white clouds in bright blue skies. Although Fragonard studied briefly with Chardin (see page 63), he was more playful and spontaneous, preferring to depict elegant and charming subjects, which he painted with grace and delicacy.

In the 1770s, the period to which this work belongs, Fragonard painted a series of portraits—of models, of his friends, of people of the theater, and of members of the aristocracy.

Fragonard has placed his sitter in profile against a gray-blue-green background. She sits comfortably propped against a large billowy pillow, absorbed in the book she holds daintily in her hand. Her left arm rests on a railing. Her auburn hair, tied with a rose-colored ribbon, sets off her lovely face with its high forehead, arched eyebrows, long lashes, small nose, beautifully shaped lips, full cheeks, and delicately curving chin. The picture enchants us with its sunshine, grace, and rhythm.

Jean-Honoré Fragonard

Jean-Honoré Fragonard was born in Grasse, France, in 1732, and spent most of his life in Paris. He worked briefly in a law office, but soon discovered his true calling. The painter with whom he studied the longest was François Boucher, who taught the young man the elegant and decorative painting style for which he is famous. After a sojourn of about five years in Italy, Fragonard returned to Paris, where he exhibited a large historical painting which won him acceptance as an associate at the French Royal Academy. After his marriage he concentrated less on depicting the pastimes of the upper classes and more on pictures describing the beauties of sentimental love, which became extremely popular. The French Revolution brought about the death, exile, or ruin of many of his wealthy patrons, however, and Fragonard died penniless in 1806.

ANTOINE WATTEAU

Italian Comedians

PIETRO LONGHI

Blindman's Buff

Antoine Watteau

Antoine Watteau was born in 1684 in the French town of Valenciennes. At the age of eighteen, he went to Paris to become a painter. After ten years of indifferent success, he was made a member of the Royal Academy. His reception painting, *Departure from the Island of Cythera*, was so unusual that an entirely new category was created for it—the *Fête Galante*, or "elegant gathering." Watteau's paintings express regret at the briefness of earthly pleasures, and are often quite poetic in feeling. Watteau died of tuberculosis in 1721.

Pietro Longhi

Born in Venice in 1702, Pietro Longhi began his artistic training with his father while still a boy. His early works were grand historical or religious paintings, but gradually he developed more intimate and delicately rendered depictions of contemporary life which earned him a considerable following. Longhi became one of the first members of the Venice Academy and taught there until shortly before his death in 1785.

In the *Italian Comedians*, Antoine Watteau has painted the curtain call of a troupe of performers of the commedia dell'arte, a type of comedy which originated in Italy in the sixteenth century. These traveling comedians performed at French estates, enacting familiar plots and making up lines to suit current events.

In the exact center of the picture, framed by the doorway in the background stage setting, stands the actor who plays Pierrot, the leading character of the play. He is dressed in a dazzling white satin costume. He has the sad, embarrassed, bashful expression of the unsuccessful suitor. A garland of roses on the steps at his feet symbolizes the approval of the audience. The leading lady, who portrays Pierrot's love, Flaminia, stands by his side. On the right are other characters in the play. In the foreground on the opposite side of the painting a jester entertains two children with the puppet Punch, lending a touch of humor to the scene. Seated above the jester are other players: Scapin strums his lute; behind him Harlequin dances. Also pictured are a romantic hero and his lady fair.

This painting was done by Watteau shortly before he died. In it he has given the viewer an insight into the parts played by the actors—roles which elicit tears, laughter, and other emotions familiar to us all.

The scene of blindman's buff as painted by Pietro Longhi is not exactly the same game which is played by young people today. In Italy in Longhi's time it was known as *pignatta*, which literally translated means "cooking pot." The object of the game was for the blindfolded player to try to break the cooking pot with the stick.

The scene takes place in a fashionable Northern Italian home. A cool, even light bathes the charming interior. The blindfolded young man holding the stick wears an embroidered gold coat, a white shirt, knee breeches, buckled shoes, and, inexplicably, a woman's transparent floral apron. The overturned brown cooking pot lies on the floor. Seated watching the young man is an older man, perhaps the father, in a long green dressing gown. The lovely young woman holding a folded fan is elegantly dressed in a white satin hooped gown trimmed with ruffles and bows of soft rose tones which echo the color of the tablecloth. She appears to be more interested in the handsome young man standing slightly behind her than in the game. Another young woman, her expression amused and intent, watches the game.

Although he worked in the traditional oil medium, Longhi has achieved a delicacy of feeling in this small painting that reminds one of the pastel pictures popular at the time.

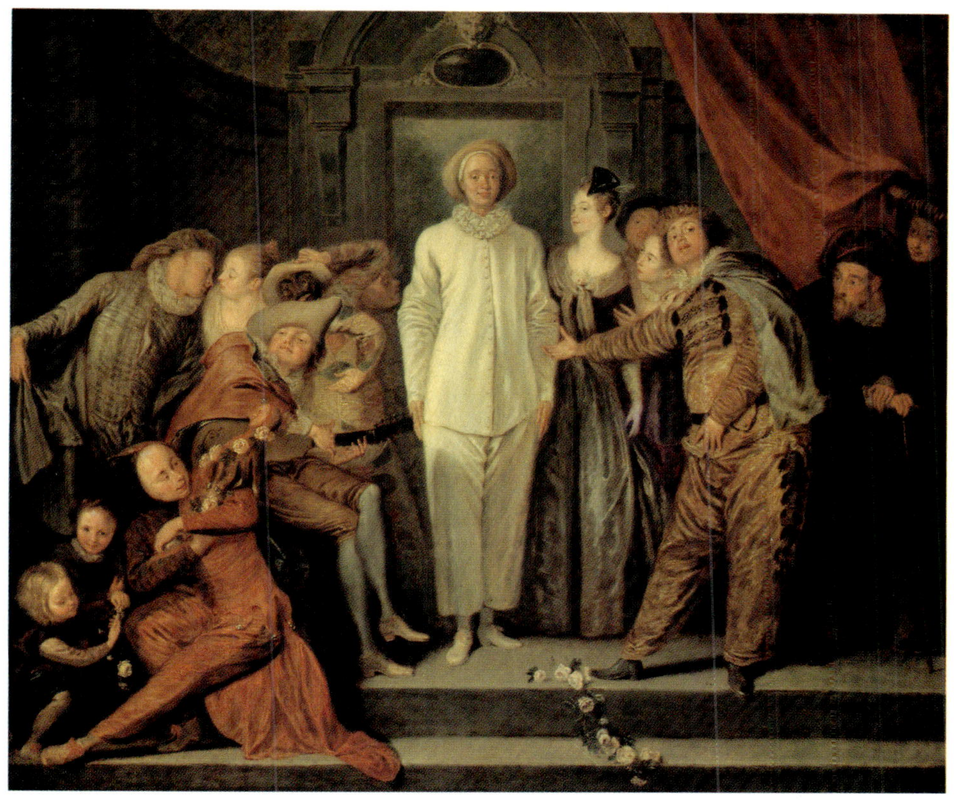

38. ANTOINE WATTEAU. *Italian Comedians.* Probably 1720. Canvas, 25 1/8 × 30″. Samuel H. Kress Collection

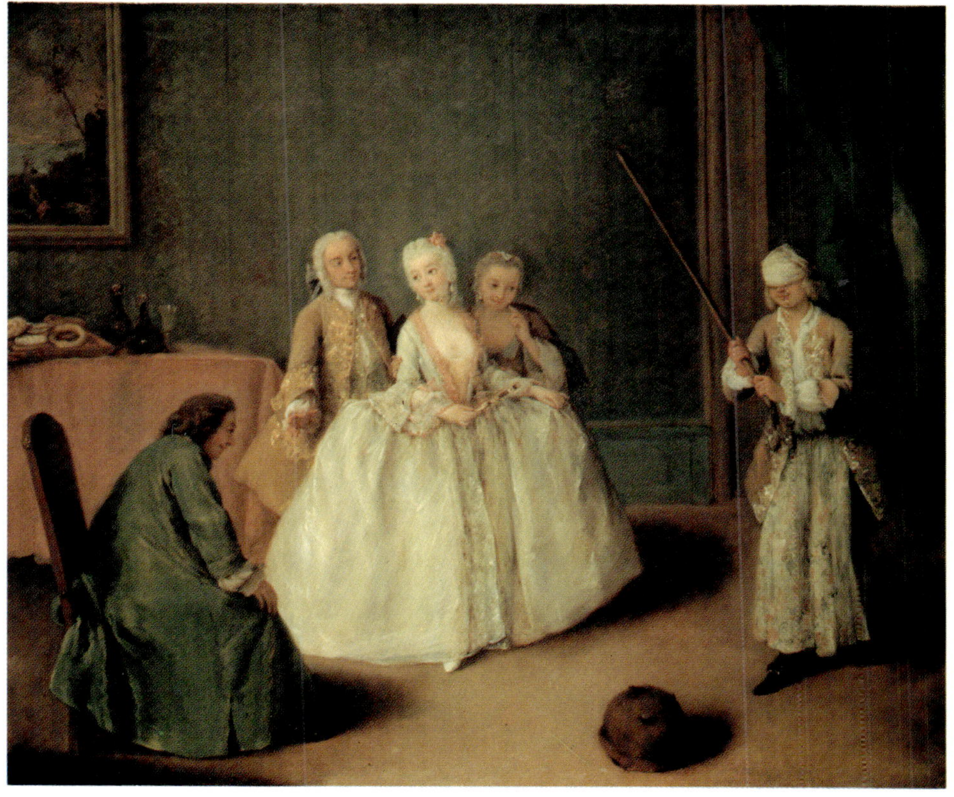

39. PIETRO LONGHI. *Blindman's Buff.* c. 1760. Canvas, 19 1/4 × 24″. Samuel H. Kress Collection

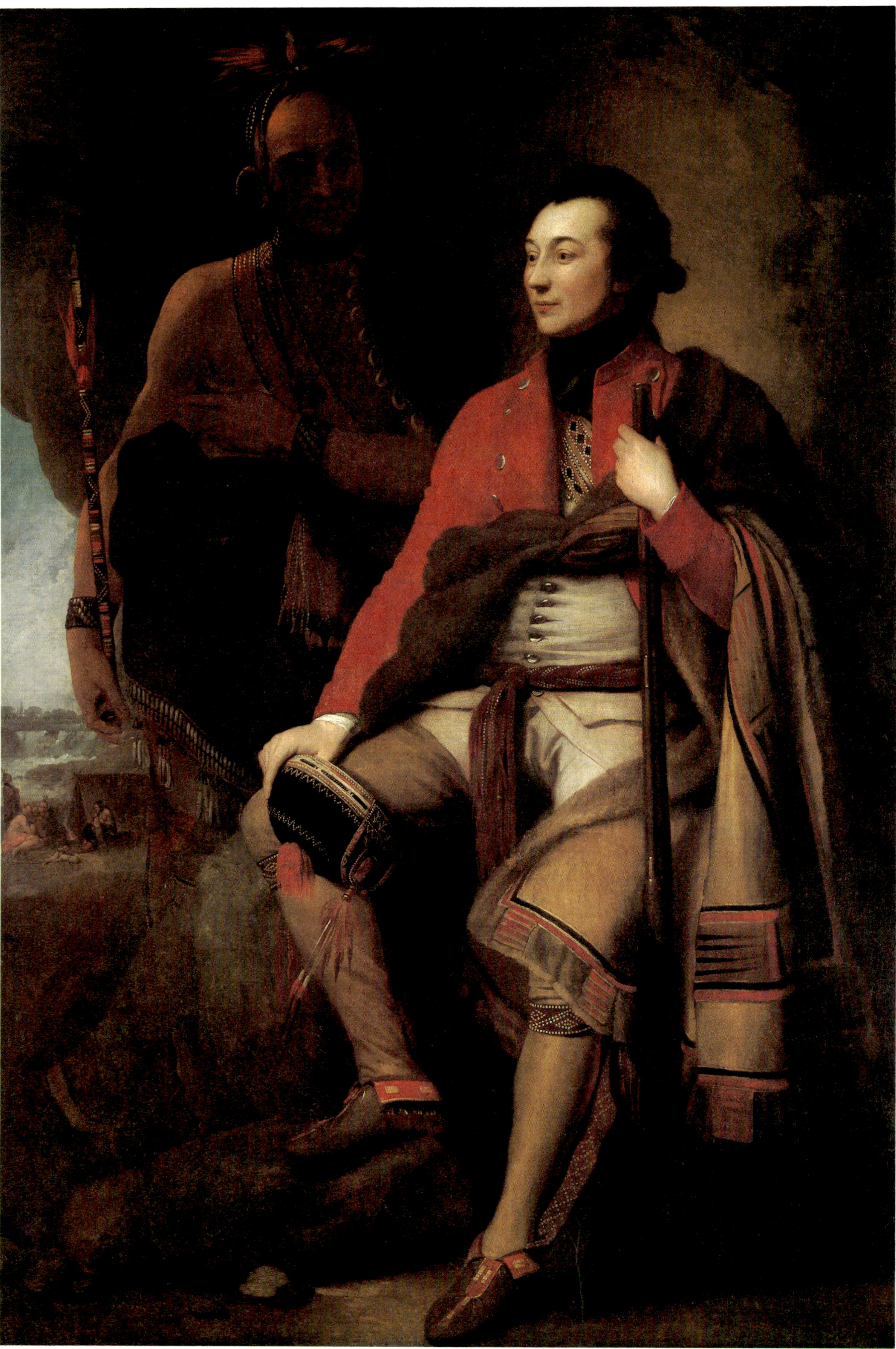

40. BENJAMIN WEST. *Colonel Guy Johnson.* 1776. Canvas, 79 3/4 × 54 1/2″. Andrew W. Mellon Collection

BENJAMIN WEST

Colonel Guy Johnson

GEORGE CATLIN

A Pawnee Warrior Sacrificing His Favorite Horse

Benjamin West's painting of Colonel Guy Johnson, the British Superintendent of Indian Affairs in the American colonies, also portrays a Mohawk chieftain, possibly Thayendanegea, known by the English name of Joseph Brant. Brant was Colonel Johnson's secretary.

The colonel is posed seated. An Indian blanket is thrown over his shoulder, and he wears Indian moccasins instead of boots. In one hand he holds a beaded black Mohawk cap with red feathers; in the other he grasps a long firearm. Across his chest and below his knees are woven bands of Indian design. His belt is adorned with *wampum*, polished shells strung in strands. West has highlighted the figure of the colonel and has further emphasized him by his red jacket and the brilliant red accents in his clothing, which make him stand out against the tans and browns which form the color scheme of the rest of the picture.

To the left of the colonel is the shadowy reddish-brown figure of the Indian chief. He holds a decorated peace pipe.

Behind the Indian, a bit of blue sky, huge waterfalls (possibly Niagara Falls), and an Indian family seated around a campfire can be seen.

In his painting West wanted to emphasize the theme of friendly cooperation between the British and the Indians. He has called attention to this theme by providing Colonel Johnson, the symbol of British authority, with some articles of Indian clothing and by giving the Indian a peace pipe. By depicting his subjects and their costumes with accuracy while at the same time stressing their dignity, the artist has created a memorable composition with an important message.

George Catlin derived his early knowledge of Indians from his mother's stories and from tales told by the hunters, explorers, Revolutionary soldiers, trappers, and Indian fighters who were

Benjamin West

The first American artist to win international fame, Benjamin West was born in Pennsylvania in 1738. By the age of eleven he was painting landscapes; a Cherokee Indian chief is said to have taught him to mix colors. Following a trip to Italy, West opened a studio in England, where he spent the rest of his life. West was the first painter to depict contemporary events with great accuracy of dress and detail, an approach which would have dramatic effects on an entire generation of artists. He was a founder of the Royal Academy in London and became its second president. American painters in particular flocked to West's studio, seeking his guidance and instruction. Benjamin West died in 1820.

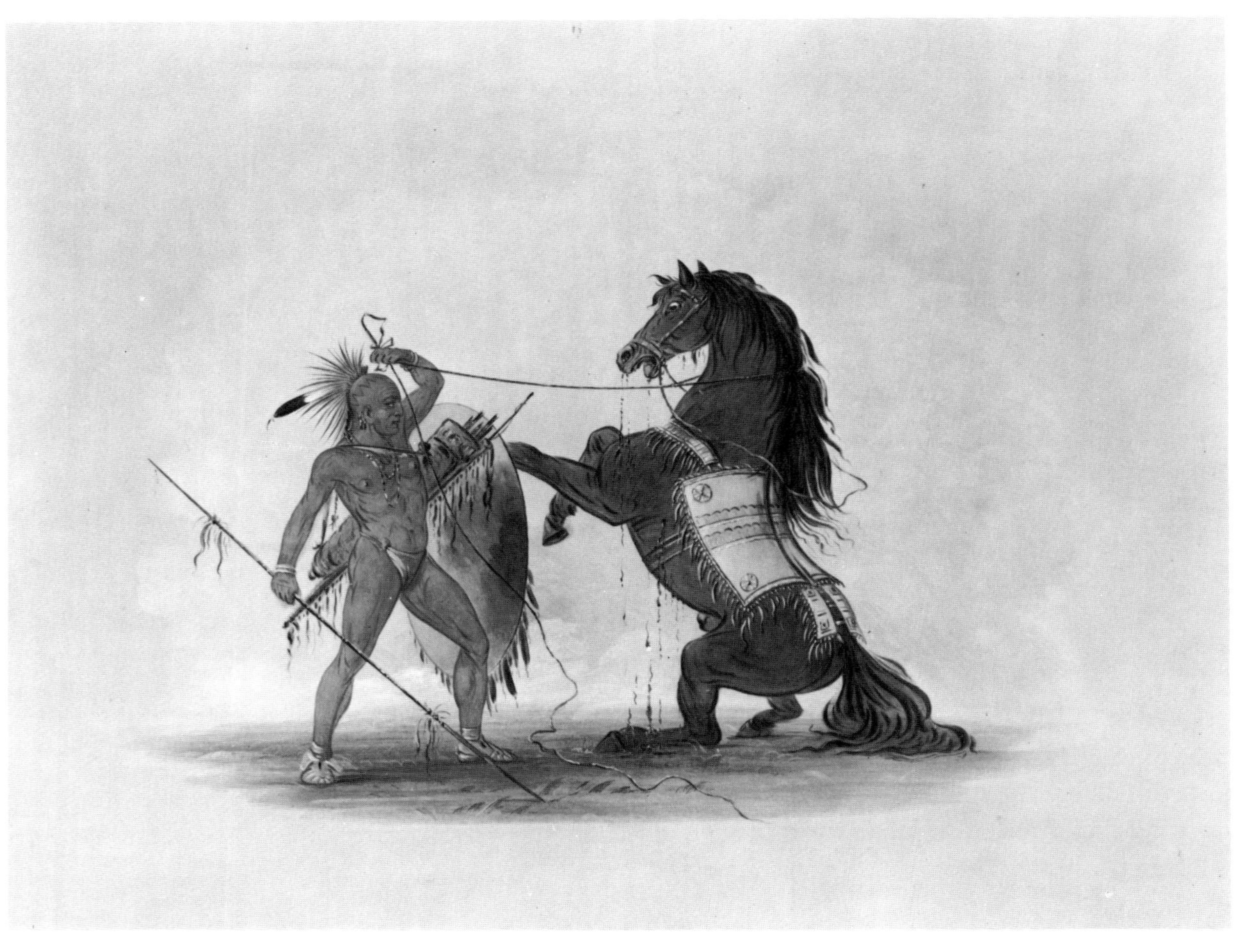

41. GEORGE CATLIN. *A Pawnee Warrior Sacrificing His Favorite Horse.* Scene observed 1832. Cardboard, 18 1/2 × 24 1/2″. Paul Mellon Collection

George Catlin

George Catlin was born in a frontier town in Pennsylvania in 1796. He received an informal education at home and then became a lawyer. When he saw a delegation of Western Indians visiting Philadelphia, however, he resolved to devote himself to the creation of a visual record of the life of the American Indian. For many years he spent the warm months painting the tribes of the Midwest and the winter months with his wife and family, painting portraits to finance his wilderness excursions. His great dream was to see his collection of Indian paintings and relics preserved in a museum, but his exhibitions, once popular, were toward the end of his life largely ignored by the public, and he was never able to raise the necessary funds for his museum. He died in 1872.

frequent guests of his family. As an adult, Catlin lived among the numerous Indian tribes of North and South America. His *A Pawnee Warrior Sacrificing His Favorite Horse*, in which a common Indian ritual is recorded, is in striking contrast to Benjamin West's portrayal of the Indian as an idealized and symbolic figure.

The event shown in the painting was observed by Catlin in Tappage Pawnee Village in 1832. According to Pawnee accounts, in the space of one day this warrior sacrificed not only his favorite horse but also seventeen other horses to the Great Spirit. Only the warrior and the Great Spirit knew the nature of the offense for which this supreme atonement was made.

The warrior, his muscular body made weightier by the large oval shield, stands out against the light background. In one hand he holds the bloodied lance; with the other he grips the horse's rein.

The horse has a frightened and bewildered expression. It rears on its hind legs, raising its front legs as if in self-protection. The mass of the large animal balances the form of the warrior and shield.

Catlin must have been greatly moved by this scene. In the most direct fashion, with great skill and feeling, he has portrayed the drama of the warrior's sacrifice.

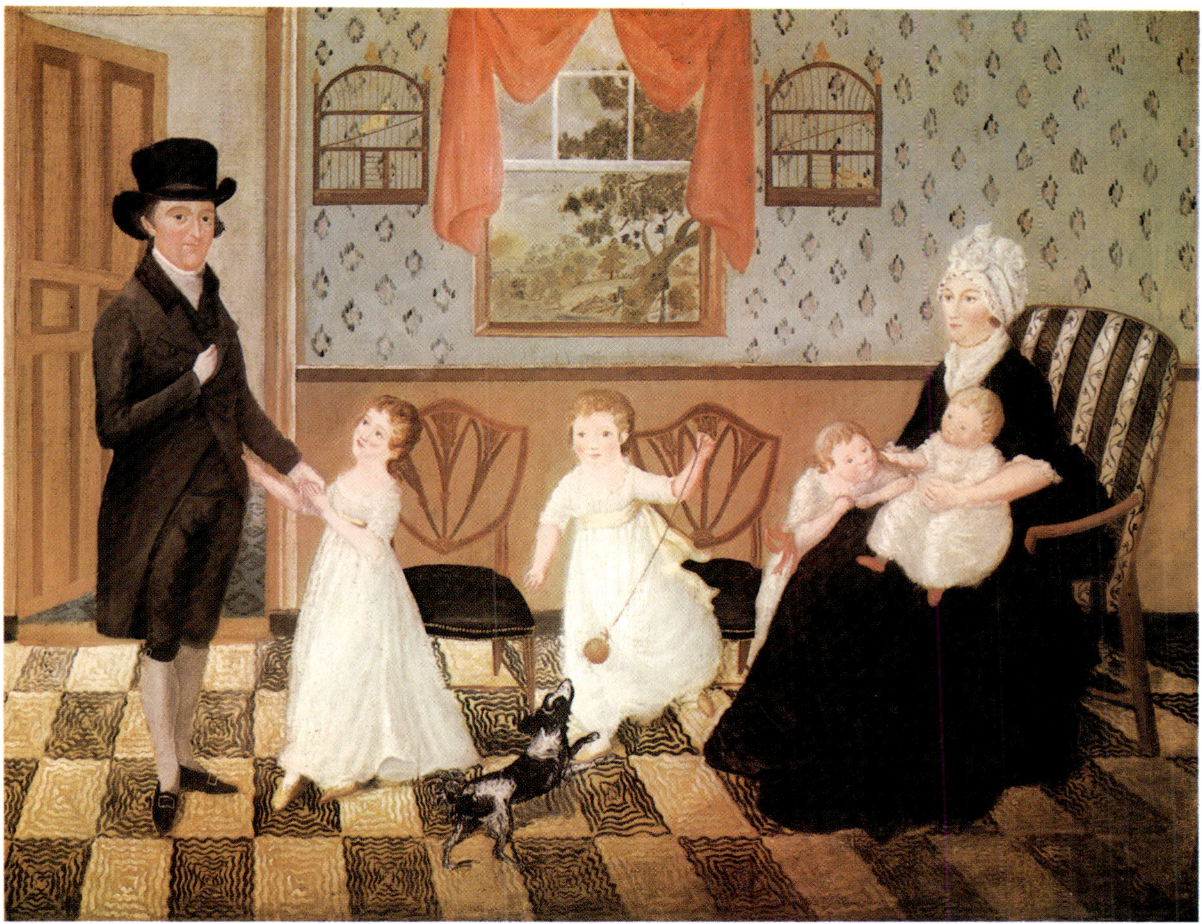

42. UNKNOWN PAINTER. *The Sargent Family*. 1800. Canvas, 38 3/8 × 50 3/8″. Gift of Edgar William and Bernice Chrysler Garbisch

UNKNOWN PAINTER

The Sargent Family

After the Revolutionary War, the United States began to grow and prosper. To express their pride in their new country and their way of life, Americans wanted pictures of themselves and their families, of their factories and farms, of the American wilderness, and of national historical events. Since they could not afford to pay the prices commanded by the professional painters who had studied in Europe, they turned to native, self-taught folk artists.

Primitive, or naive, painters, as they are called, had no formal training in art. Usually they showed forms in a flat way, for they had not studied anatomy or learned how light and shade can be used to create the illusion of three-dimensionality. Their pictures, similar to those of children, are decorative and charming.

As with many other works of American folk art, we do not know the name of the artist who painted this New England family group.

The figures, especially the baby, have a rather stiff, doll-like quality which enhances their appeal. The clothes and furnishings reflect the taste of the American Federal period. The checkered floor, the figured wallpaper, and the bird cages on either side of the draped window form a sprightly patterned setting for the figures. The window serves as a picture frame for the wooded landscape beyond.

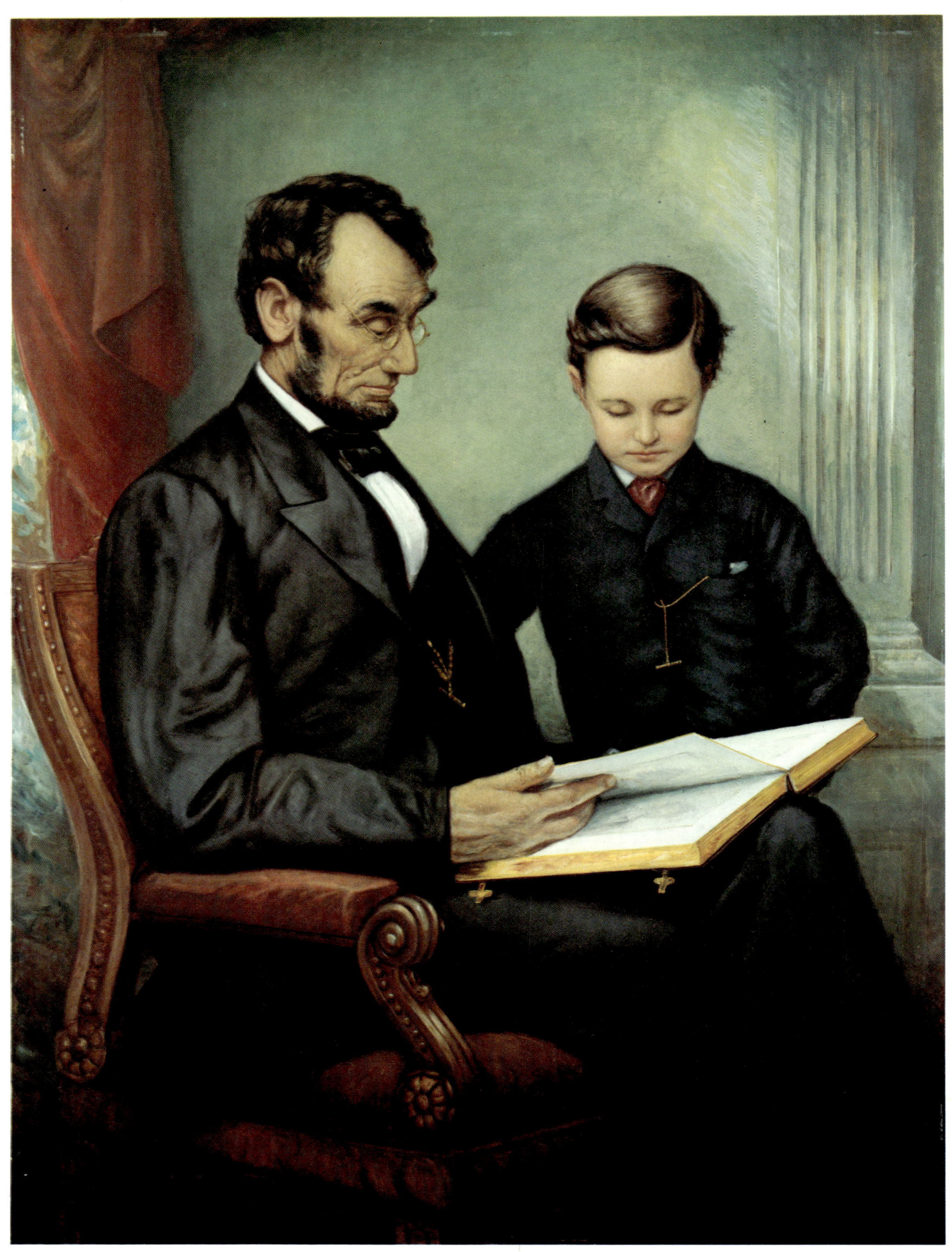

43. FRANKLIN COURTER. *Lincoln and His Son Tad.* c. 1929. Wood, 46 × 35 1/2″. Andrew W. Mellon Collection

FRANKLIN COURTER

Lincoln and His Son Tad

Painted about 1929, Franklin Courter's picture of Abraham Lincoln and his beloved Tad was based on a nineteenth-century photograph. While waiting to have his picture taken in the studio of the famous Washington photographer Matthew B. Brady, Mr. Lincoln picked up an album of Brady's pictures and began showing them to his son. A partner of Brady's, Anthony Berger, noticed the pair by chance and was so impressed by their pose that he asked them to remain as they were so that he could take their picture.

Tad, or Thomas, the fourth of the Lincoln sons, was born on April 4, 1853. He was a lovable, warmhearted, responsive, and devoted child whose escapades in the White House were the talk of the nation, though often a trial to some of the Cabinet members, office seekers, and other visitors to the White House. He once shattered the decorum of a reception in the East Room by driving his pet goat into the midst of the guests. Yet he inherited his father's sensitive understanding of the feelings of others: it is said that he took the President to meet a lame boy whose father had been killed in the Civil War so that the child could have the thrill of meeting Lincoln.

In the painting we are given a profile view of the rugged, war-weary face of Lincoln. His somewhat curly dark hair is flecked with gray, as are his eyebrows and the beard which frames his thoughtful face. He wears small gold-rimmed glasses; a gold watch chain stands out clearly against his black vest and suit.

Tad stands near his father, looking down at the gilt-edged photograph album with an absorbed expression. His dark eyebrows and lowered eyes are in striking contrast to the delicate tones of his face. He wears a thin gold watch chain similar to the one worn by the President.

When Tad was eleven he went with Lincoln to see General Grant at the time Richmond was taken; together they toured the fallen city. It was their last happy excursion together. Shortly thereafter the President was shot.

After his father's death, Tad was a devoted son to his saddened mother and a great comfort to her. For several years he traveled with her in Europe, but when they returned to America during the early months of 1871, Tad was ill. On July 15 of that year, he died at the untimely age of eighteen.

Franklin Courter

Born in Caldwell, New Jersey, in 1854, Franklin Courter became a professor of drawing and painting at Albion College in Michigan. It was there that he painted his first portrait of Abraham Lincoln and began his lifelong study of the former President. For over forty years, Courter studied all available material on Lincoln and sought the descriptions of those who had known or seen him, until he was so familiar with his subject that his portraits of Lincoln could almost be thought to have been painted from life. Courter died about 1935.

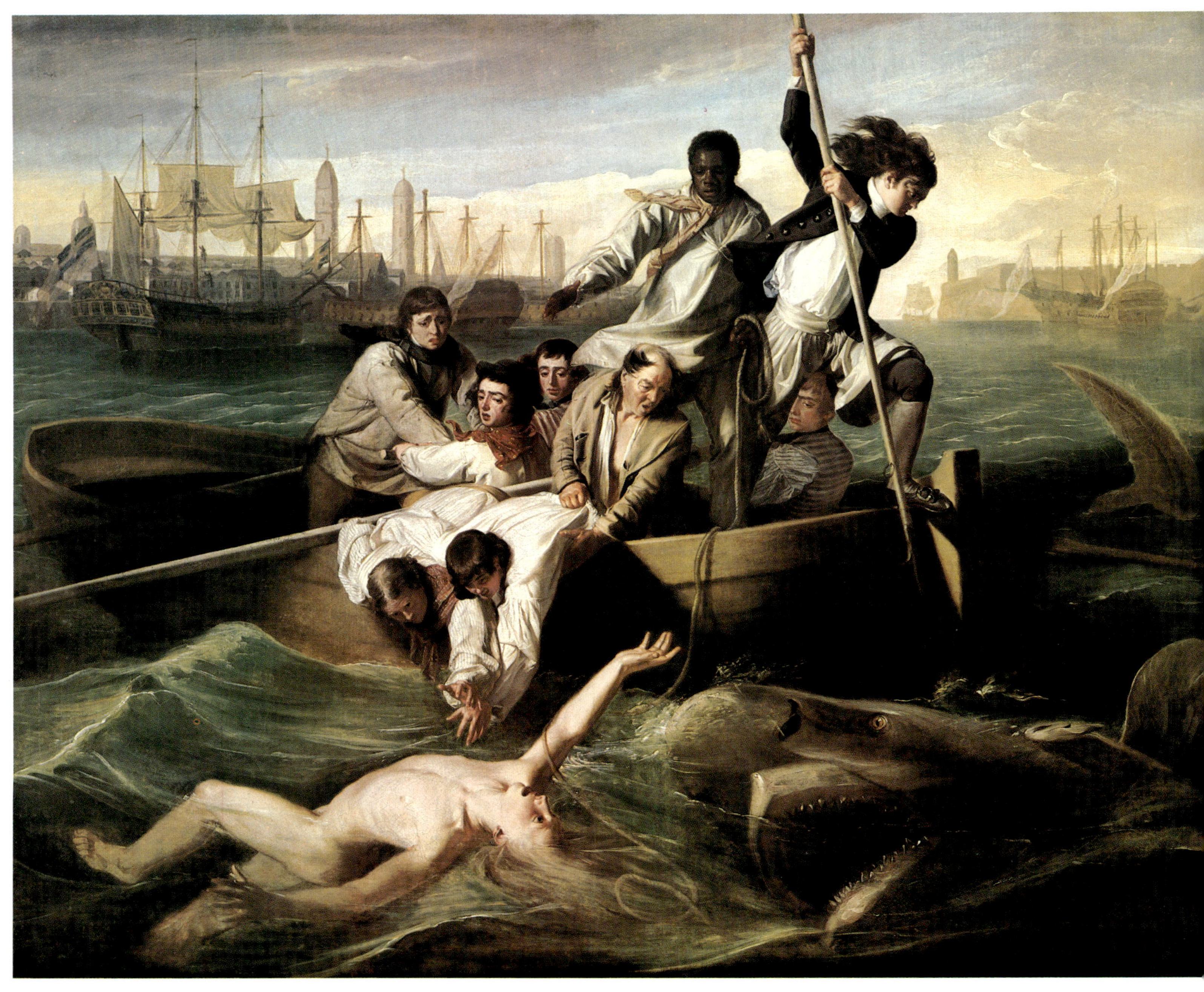

44. JOHN SINGLETON COPLEY. *Watson and the Shark*. 1778. Canvas, 71 3/4 × 90 1/2″. Ferdinand Lammot Belin Fund

John Singleton Copley

Born in or near Boston in 1738, John Singleton Copley was one of America's first great artists. His stepfather was an engraver and painter who helped inspire him at an early age to become an artist. By the age of twenty-five, Copley was already one of the leading portraitists in the Colonies, but he longed to learn the grand painting style of Europe. After his marriage, he traveled through France and Italy, finally settling in London. There he soon distinguished himself with dramatic paintings of recent historical events and elegant portraits, and became a member of the Royal Academy. Copley died in London in 1815.

JOHN SINGLETON COPLEY

Watson and the Shark

The artist John Singleton Copley has in this painting shown an incident which actually happened.

In 1749 fourteen-year-old Brook Watson was a sailor on a ship that had docked in the harbor of Havana, Cuba. It was very hot, and Watson decided to cool off by going for a swim. What started out to be a lark for the young sailor turned into a disaster. A shark attacked him and bit off his foot. The monster returned again for a second attack.

In the dark brown boat are four sailors. Their faces express their fears. Two lads lean over the edge of the boat, trying to reach the struggling victim. An old tar grasps the shirt of one of the boys to prevent him from falling overboard. At the same time he keeps an anxious eye on the approaching shark. The West Indian, his head outlined against the white clouds and blue sky, has thrown a rope to Watson, but the boy has failed to catch it. The man on the right, dark hair flowing in the wind, stands firmly balanced with his foot on the dory's bow, trying to dispose of the shark with a boat hook.

Copley presents the scene with great accuracy. Young Watson is partially submerged in the clear green-blue water. His right leg is underwater, hiding the dreadful wound. His head is thrown back; his open mouth, rolled eye, and outstretched arm desperately reaching for help express his terror. The drama is underscored by the vivid contrasts of color and of textures: the boy's flesh, the skin of the fish, the turbulent water—all are painstakingly rendered. The curving gray shark is dangerously close to Watson; its gleaming eye is fixed on the struggling boy, its large jaw open, exposing sharp jagged teeth.

In contrast to the gruesome scene in the foreground is the calm Cuban harbor. Etched against the sky are the cathedral's dome, two convent towers, and on the right, El Morro Castle. At anchor on the left are a number of square-rigged sailing vessels.

Will Watson be devoured by the shark or will the boat hook kill the bloodthirsty beast? Copley successfully captures this suspense in all its melodramatic detail. Actually, there was a happy ending to the adventure. Watson was rescued. He lived to become lord mayor of London and tell Copley the story.

Though the fashion of the time was to paint dramatic pictures based on the acts of heroic figures of literature or history, in this exciting painting Copley dared to show the heroism of ordinary people of his own time.

FRANCISCO JOSÉ DE GOYA

Condesa de Chinchón

Victor Guye

The Condesa de Chinchón, María Teresa de Borbón y Vallabriga, was a niece of King Charles IV of Spain. Francisco Goya became a close friend of the royal family, and while visiting at their country estate near Avila he painted this attractive portrait.

An inscription in the lower-left-hand corner of the picture tells us that María Teresa was two years and nine months old when the portrait was painted. The quaint child stands self-assured in front of a low gray wall against a background of green hills topped by majestic blue-gray mountains. The child's small figure and delicate face, wide-open eyes, and rosebud mouth are in striking contrast to the grown-up clothes she wears. Goya's treatment of María Teresa's costume is highly realistic: her beautiful blue silk bodice is edged with ruffles of lace; her long full dark-blue skirt reaches nearly to the ground but permits a view of a lighter blue satin shoe decorated with a silver buckle. Seated near the child is a fluffy white dog. There is a touch of wistfulness in her face which arouses our sympathy.

Victor Guye was the tousle-haired court page of Joseph Bonaparte, who became ruler of French-occupied Spain in 1808.

Looking at his features, we can see that he is not a conventionally handsome boy. The eyes are too far apart, the nose too large and blunt, the mouth too small. Yet what an appeal there is in the face that stares out so unhappily from the canvas! The wide eyes plead for understanding, and the boy's mouth reflects his anxiety. Young Victor is clad in black and weighed down under a mass of gold braid. Mechanically, as if in obedience to a command, he holds an open book. To emphasize his loneliness, Goya has placed him in an indefinite space against a deep red-brown darkness—alone but for his shadow. It is characteristic of Goya that he depicts the features of this small boy with such realism, and just as characteristic that he tells us so much about Victor's inner feelings. There is a sense of compassion here that is absent from many of Goya's portraits.

Francisco José de Goya

Francisco José de Goya was born in the province of Saragossa, Spain, in 1746. He began painting at an early age, and eventually married his teacher's sister. Among Goya's first commissioned works were designs for tapestries, which led to many honors. Once he became known, Goya was elected president of the Royal Academy and became First Court Painter to King Charles IV. He later served the French conquerors of Spain, but at the same time made a long series of etchings condemning the atrocities and misery that had followed the French invasion. When Goya found the restored Spanish monarchy to be just as repressive as the French had been, he went into voluntary exile in Bordeaux, France. Though left stone-deaf by an illness in 1792, Goya continued to paint actively until his death in Bordeaux in 1828.

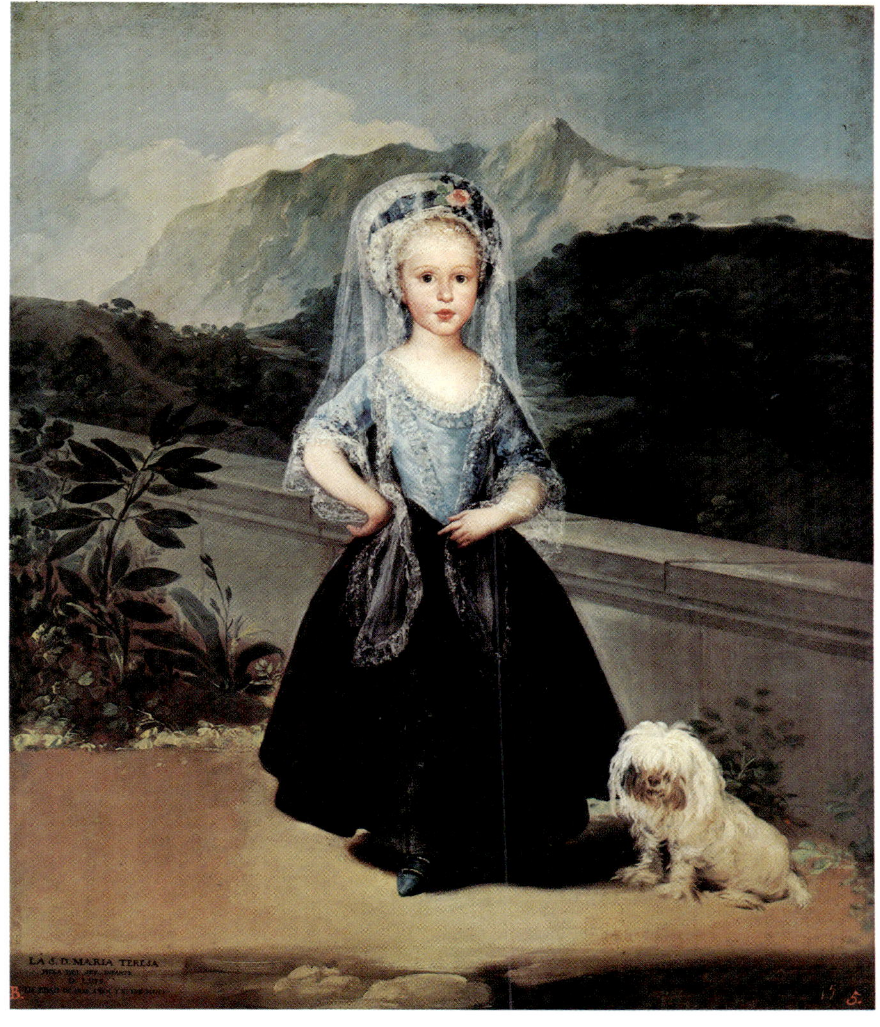

45. FRANCISCO JOSÉ DE GOYA. *Condesa de Chinchón*. 1783. Canvas, 53 × 46 1/2″. Ailsa Mellon Bruce Collection

46. FRANCISCO JOSÉ DE GOYA. *Victor Guye*. 1810. Canvas, 42 × 33 1/2″. Gift of William Nelson Cromwell ▷

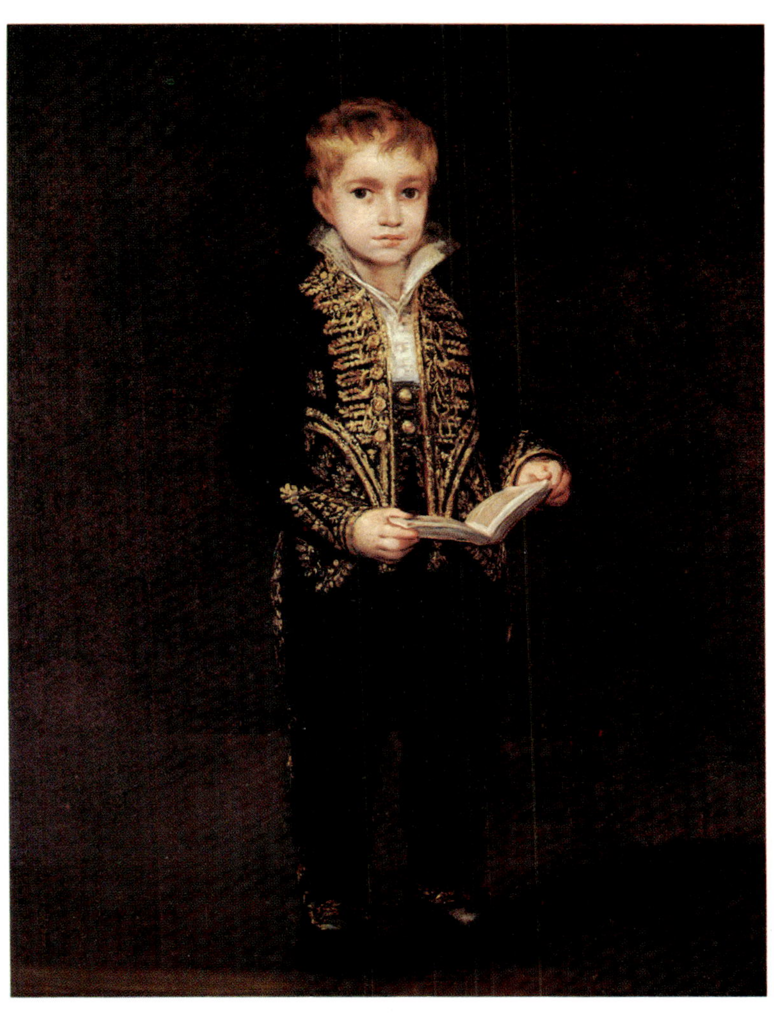

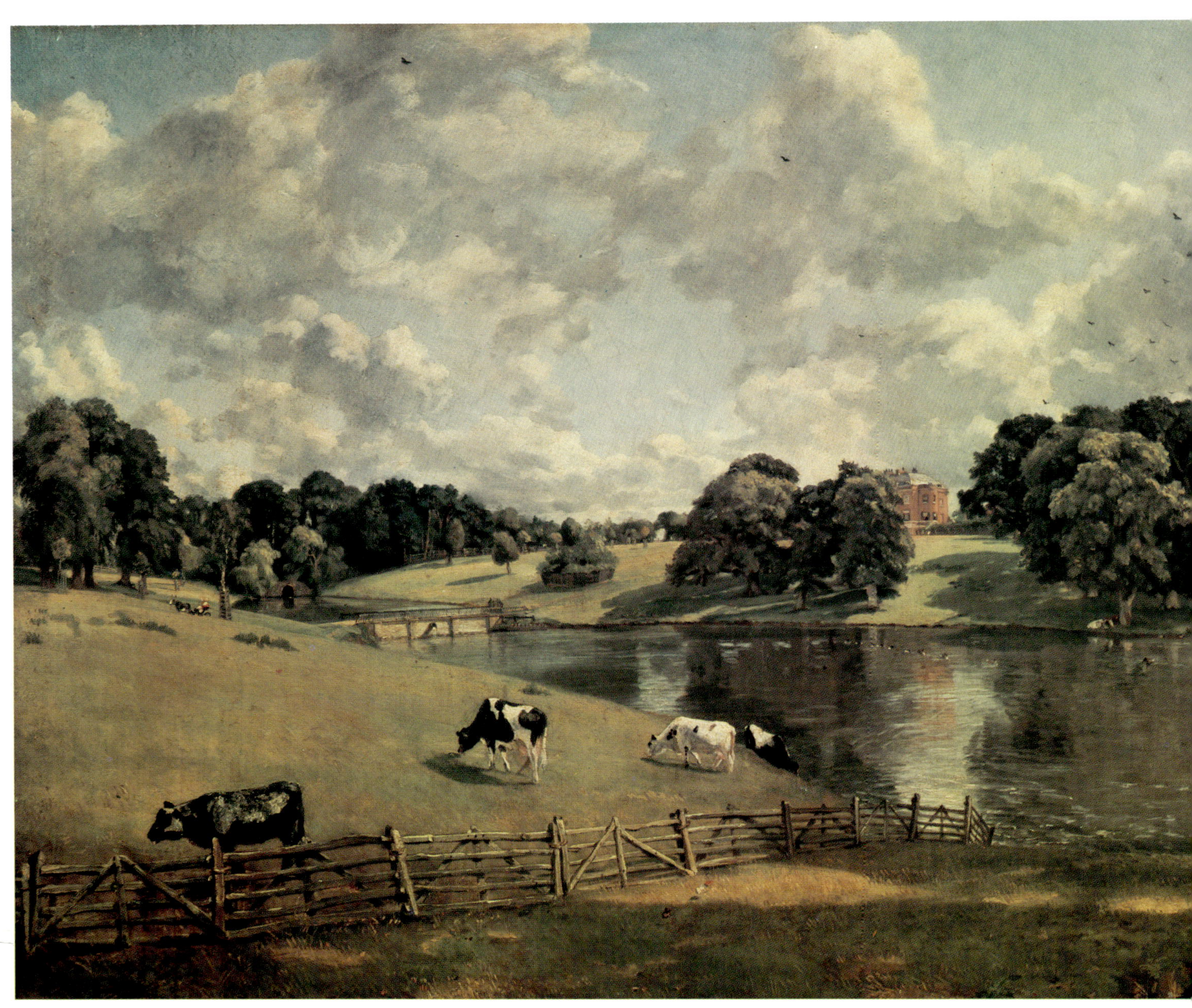

47. JOHN CONSTABLE. *Wivenhoe Park, Essex*. 1816. Canvas, 22 1/8 × 39 7/8″. Widener Collection

JOHN CONSTABLE

Wivenhoe Park, Essex

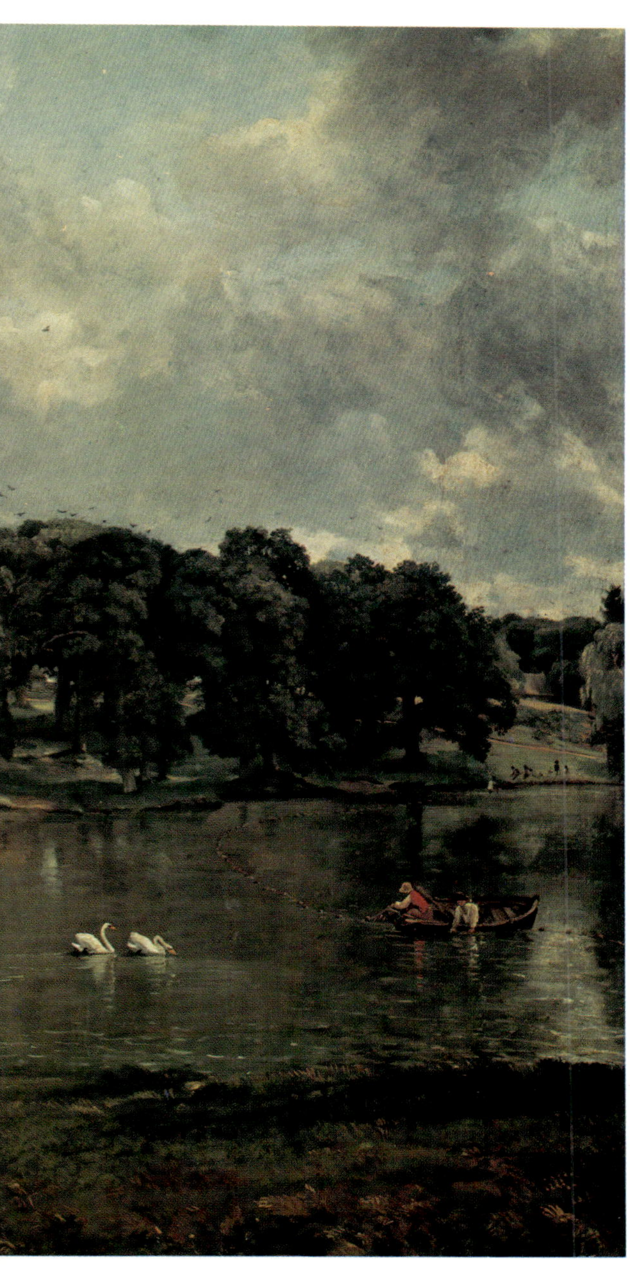

In 1816, while visiting Wivenhoe Park, the country estate of Major-General Francis Rebow, John Constable was commissioned to paint a view of the general's house and the park area surrounding it. Unlike other painters of the time, who usually painted in their studios from sketches, Constable here worked directly from the scene.

In the center of the picture amid a clearing in the trees is a rose-red brick house with a gray-white roof. There are a number of chimneys, indicating that the house is a large one. In his rendition of the landscape, the artist has conveyed his love of nature and its changing moods. The blue-gray sky is darkened by great billowy clouds. Spots of sunlight and shadow are scattered across the clumps of green trees, the lake, and the pasture in the foreground of the painting. Dark flecks in the sky represent flying birds.

The house is reflected in the shimmering lake, where two graceful white swans and a line of ducks are swimming. Near the swans are two small figures fishing from a boat. Our attention is drawn to them by a touch of red in the clothing of one of the figures.

A number of black-and-white cows graze in the sun on the grassy pasture. Beyond and barely visible is a donkey cart driven by the general's daughter. Here again the artist accents the figure with a tiny dab of red.

Constable began this as a smaller work, but because the Rebows wanted to include the deer house, whose roof can be seen in the clearing beyond the hill, he had to add a three-inch strip of canvas to the right side of the painting. So that the main house would still be at the center, he also added a strip at the left. He took advantage of this additional space to include the donkey cart.

It is a tribute to Constable's art that his painting is not only a record of a country estate, but a delightful pastoral scene enriched by sparkling lights and dramatic shadows and painted in colors that vividly bring to life the soft, dewy hills and windy skies of his beloved England.

John Constable

One of England's most famous landscape painters, John Constable was born in Suffolk, England, in 1776. He worked for a time in his father's mill, but his drawings soon came to the attention of a local gentleman, who sent him to London to study. At first he had no success and returned to the mill, but eventually his pictures began to be accepted for exhibition. He was then able to live comfortably as an artist in the countryside he loved so much to paint. He was elected a member of the Royal Academy, and was preparing a painting for exhibition there when he died in 1837.

JOSEPH MALLORD WILLIAM TURNER

Mortlake Terrace

About ten years after Constable painted *Wivenhoe Park, Essex,* Joseph Mallord William Turner painted *Mortlake Terrace.* Both of these British artists had a strong impact on landscape painting, in Turner's case helping to change it from a precise pictorial record of detailed fact to a more impressionistic interpretation of the scene.

Mortlake Terrace depicts a tree-lined garden on the banks of the Thames River. In the background on the left are a large summerhouse and some village buildings. On the river the ceremonial barge of the lord mayor leads a procession of boats. In the foreground and middle ground are various objects—a hoop leaning against a wall, a ladder, a table, a chair, and some abandoned toys. By the wall people watch the passing boats. Standing on the wall, serving as the focal point of the picture, is a dog, who appears to be barking at the boats. The dog was not in the original painting, but is a paper cutout which Edwin Landseer, the famous English animal painter, stuck on the canvas during Turner's absence. Turner liked the addition, but moved it a fraction and painted it black. The dog has remained there ever since.

Touching everything with luminous light is the glowing setting sun, whose rays glimmer across the water and pierce the brown wall, seeming to cause it to dissolve at one point. So brilliant is the sunlight that it washes out the natural colors of the trees, garden walk, foliage and grass, buildings and boats, while at the same time unifying the objects in a golden haze. Thus, the scene itself is of secondary importance, for the light-filled atmosphere dominates everything, exactly according to the artist's intentions.

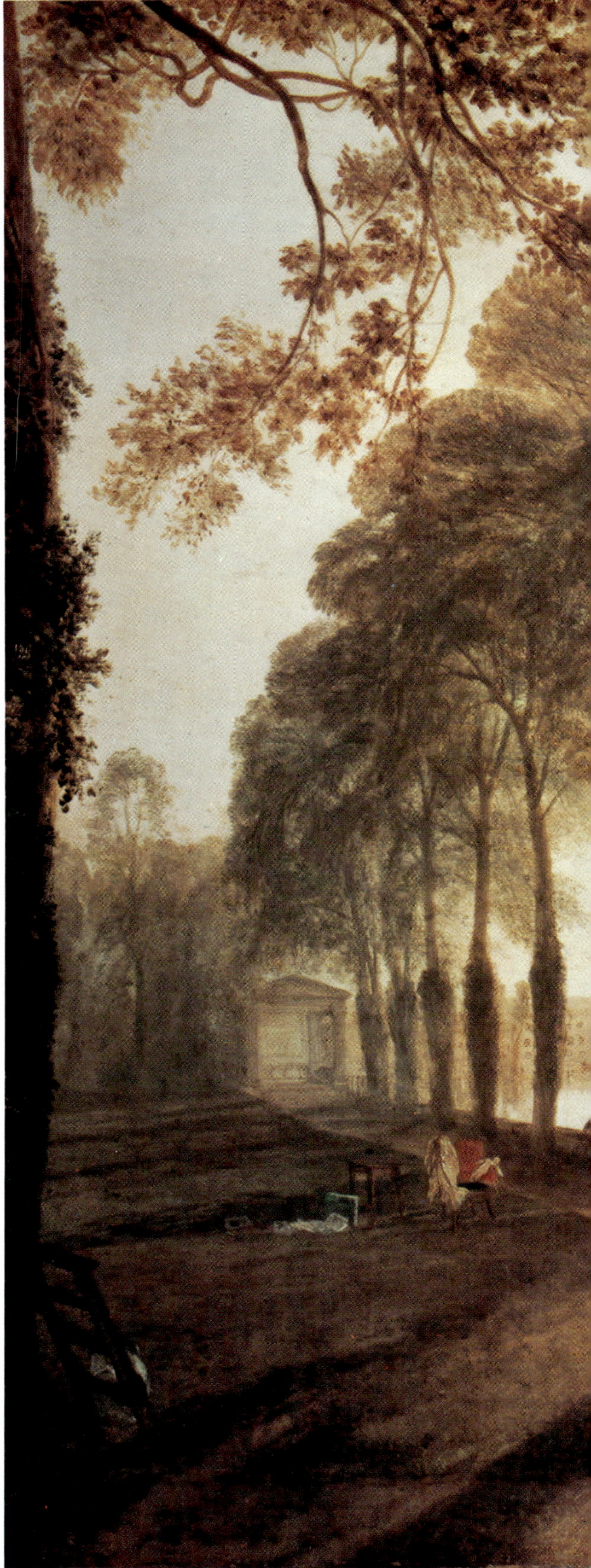

Joseph Mallord William Turner

The son of a barber, Joseph Mallord William Turner was born in 1775 in London. As a boy he became skilled at watercolors and drawings of picturesque scenes. Turner had his own studio by the age of eighteen and became one of the youngest artists to be granted full membership in the Royal Academy. His landscapes, grand in conception and dazzling in color, were unanimously praised. Later in his career, Turner's pictures became increasingly abstract, often representing little more than color, light, and tumultuous motion. Because the subject was less recognizable, these paintings were not as popular as Turner's more youthful work, but today they have become the most admired of all. Turner died in 1851.

48. JOSEPH MALLORD WILLIAM TURNER. *Mortlake Terrace*. c. 1826. Canvas, 36 1/4 × 48 1/8″. Andrew W. Mellon Collection

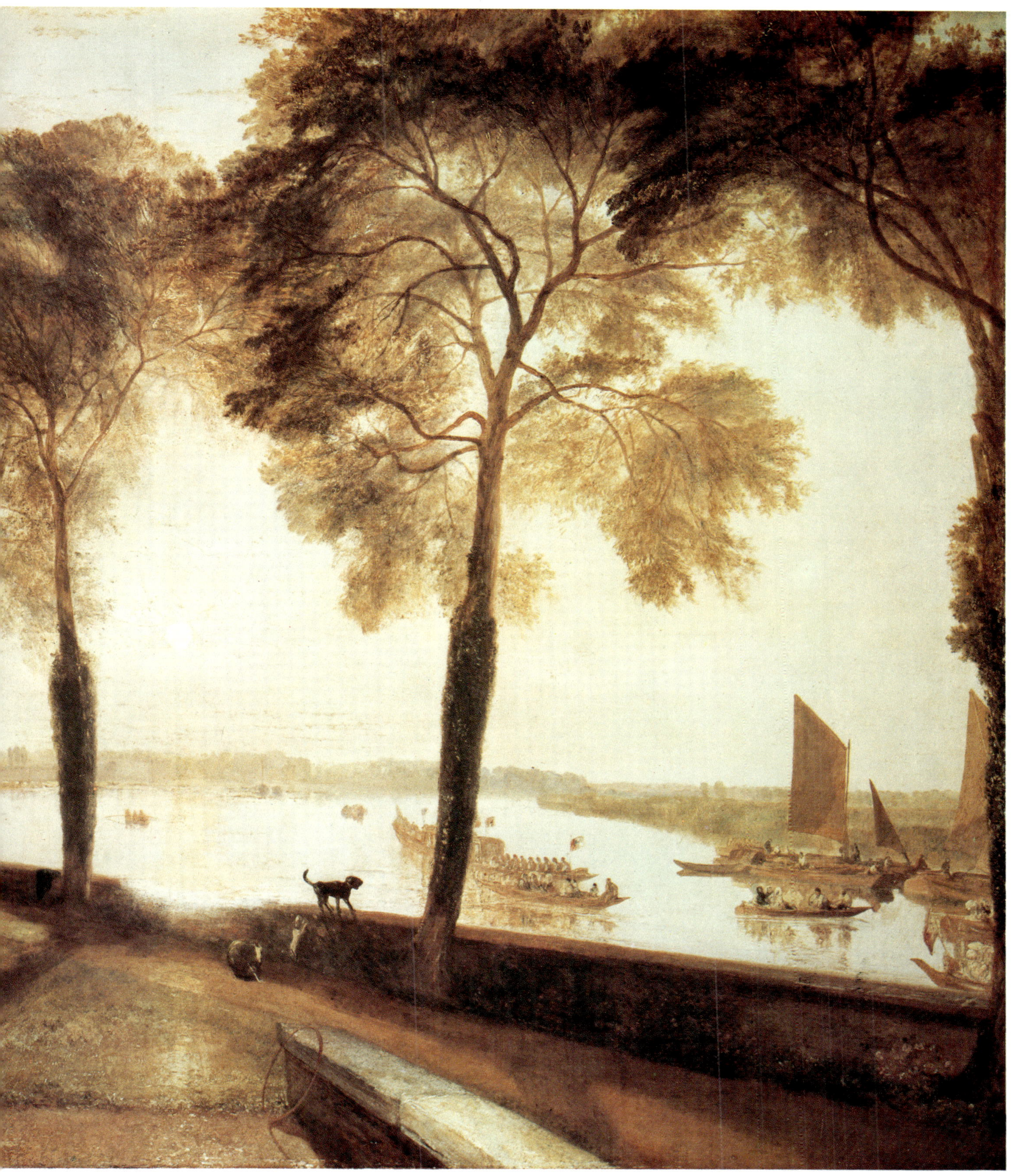

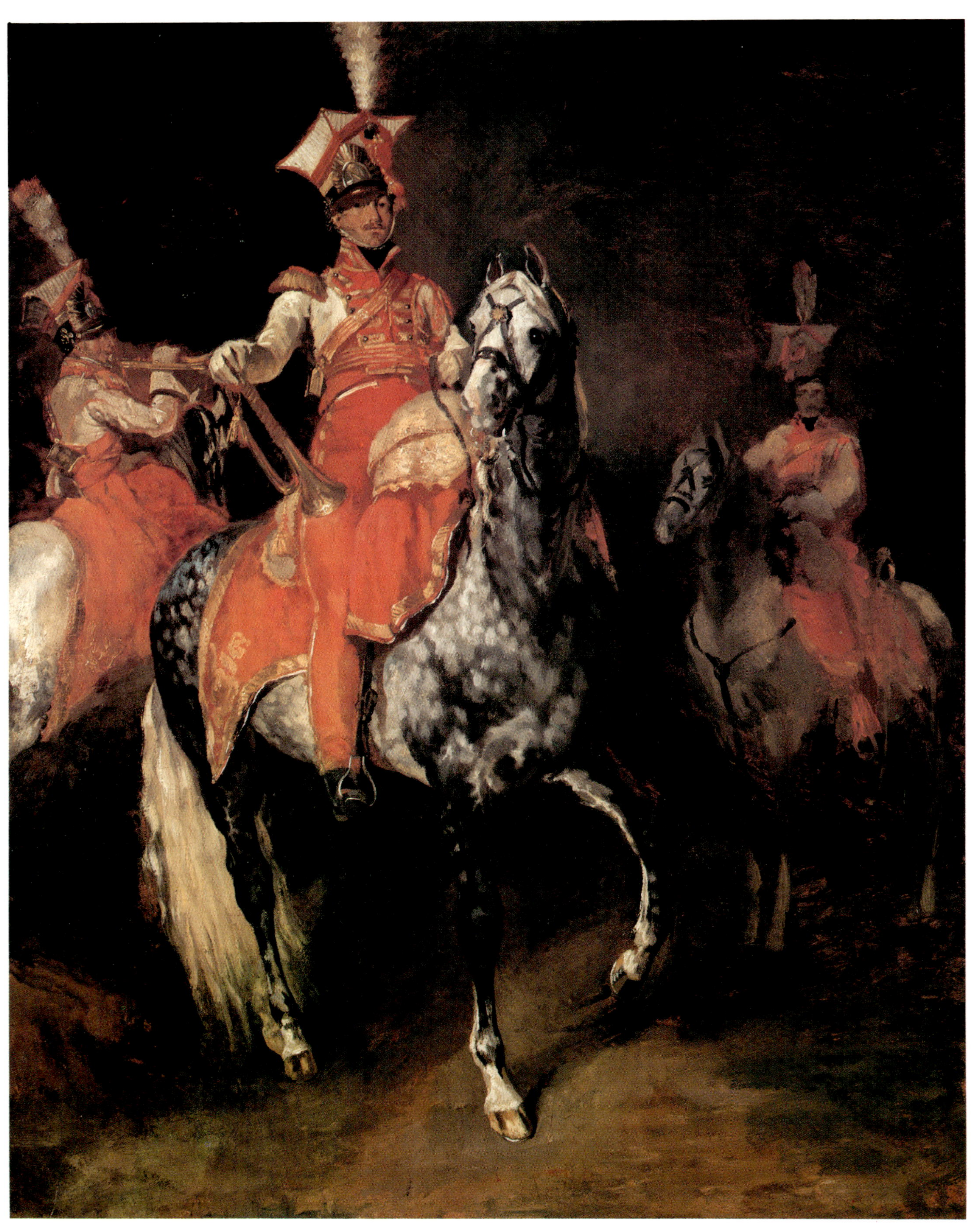

49. THÉODORE GÉRICAULT. *Trumpeters of Napoleon's Imperial Guard.* 1812/14. Canvas, 23 3/4 × 19 1/2″. Chester Dale Fund

THÉODORE GÉRICAULT

Trumpeters of Napoleon's Imperial Guard

In *Trumpeters of Napoleon's Imperial Guard,* Théodore Géricault has arranged his composition to suggest that it is a small portion of a larger theater of action.

Géricault has given the painting a feeling of urgency and vitality through the use of fresh, vivid colors and dramatic lighting.

In the center, astride a large gray dappled horse covered with a red robe, is a trumpeter of the Second Regiment of Napoleon's Imperial Guard.

The trumpeter's uniform is vivid red trimmed with bright gold and white. Recruited from Poland, like most members of the regiment, he wears the Polish Lancer's white square-top *czapka* (cap) edged with red and crowned with a large white plume touched with red. The Napoleonic *N* is emblazoned on a brass sunburst plate at the front of the cap. In one hand he holds his gleaming trumpet; with the other he grasps the reins of the horse. He appears to be listening for the call to battle.

In the background on the left, seated on a white horse turned in an opposing direction to the central figure, is a second trumpeter, who sounds the call to action. The light coming from the left focuses attention on the two figures. The highlights dance over their uniforms, visors, and trumpets.

The dimly seen figure on the right contrasts with the two brilliant trumpeters and helps to create a feeling of depth in the painting.

Géricault was a founder of Romanticism, an early-nineteenth-century movement that emphasized strong, even violent, moods. The dashing soldiers on spirited horses in this picture are typically Romantic subjects, while the bold contrast of red and gold against the dark, shadowy background demonstrates the gusto of the Romantic style.

Théodore Géricault

Born in Rouen, France, in 1791, Théodore Géricault moved to Paris with his wealthy family as a small boy. From the age of sixteen he used the living allowance he received from his father to study art and soon acquired a dashing and dramatic style. Géricault painted subjects that appealed to his romantic nature—horse races, military officers, and storm or battle scenes. His most famous picture, *The Raft of the Medusa,* was inspired by a tragic shipwreck. The emotional quality of Géricault's art parallels his life. He enjoyed riding spirited horses, and a series of racing accidents contributed to his early death in 1824.

EDOUARD MANET

The Old Musician

The Dead Toreador

Edouard Manet

Edouard Manet was born in Paris in 1832. He was a notoriously poor student and at eighteen resolved to become an artist despite the objections of his parents. After more than ten years of study, travel, and very little public recognition, Manet exhibited a picture entitled *Luncheon on the Grass* at the newly established Salon des Refusés, or "salon of the rejected." Because it showed a nude woman with two dressed men in broad daylight and was painted in an unconventional broad, flat manner, the picture caused a great scandal. Manet's work continued to be condemned or misunderstood until late in his life. Today he is generally considered to be the first painter of the modern age. Edouard Manet was awarded the Legion of Honor the year before his death in 1883.

The shabbily dressed figures in Manet's *The Old Musician* have a frozen quality, as if they had suddenly stopped whatever activities they were engaged in to be photographed. Manet strengthens this effect by using a bright light which flattens the forms and makes us aware that they are composed of separate shapes of color. The central figure is a musician, a picturesque character from the bohemian quarter of Paris. Unkempt though he is, he has a sense of dignity. His cloak is thrown over his shoulder like a Roman toga. He gazes at the spectator, his fingers absently plucking a melody on his violin. Behind him is a top-hatted figure wrapped in a blanket, his face curiously out of focus. The model for this figure was a rag-picker who frequented the galleries of the Louvre. Beside him, partly cut off by the frame, is a turbaned Oriental in a long robe; this mysterious figure seems rapt in thought. To the left Manet has shown two boys from the street and a ragged barefoot girl holding a baby. The people seem isolated in their own private worlds. The picture has an air of unreality that sets it apart from all but a few of Manet's works.

In *The Dead Toreador* Edouard Manet has portrayed his subject against a somber background. There is no gore, no crowds, no bull, no ring. Yet the small spots of red on the bullfighter's white cravat and on the ground near his shoulder, the pallid mauve-pink cape still clutched in his hand, the hilt of the sword near his right shoulder, and his black silk costume are effectively used by Manet to convince us that the man is dead.

The painting was originally part of a larger composition which showed the toreador lying diagonally in the foreground of the work, a bull stomping the ground in the center, and three *banderilleros* (whose role in a bullfight is to implant a decorated barbed dart into the bull's neck or shoulder muscles) standing against a fence.

When the painting was exhibited in 1864 at the Salon, the official exhibition of the French Academy of Painting and Sculpture, it met with unfavorable criticism. The leaders of the French art world, and the public as well, were not ready to accept the realism of Manet's work. Not only were they horrified by the subject matter, they were upset by the painting's flat shapes and strong color contrasts. It was totally different from the sleek works by popular artists who usually chose heroic scenes from history or mythology as their themes.

Manet reacted by taking the canvas back and cutting it apart. Another portion of the work is called *The Bullfight* and hangs in the Frick Collection in New York City.

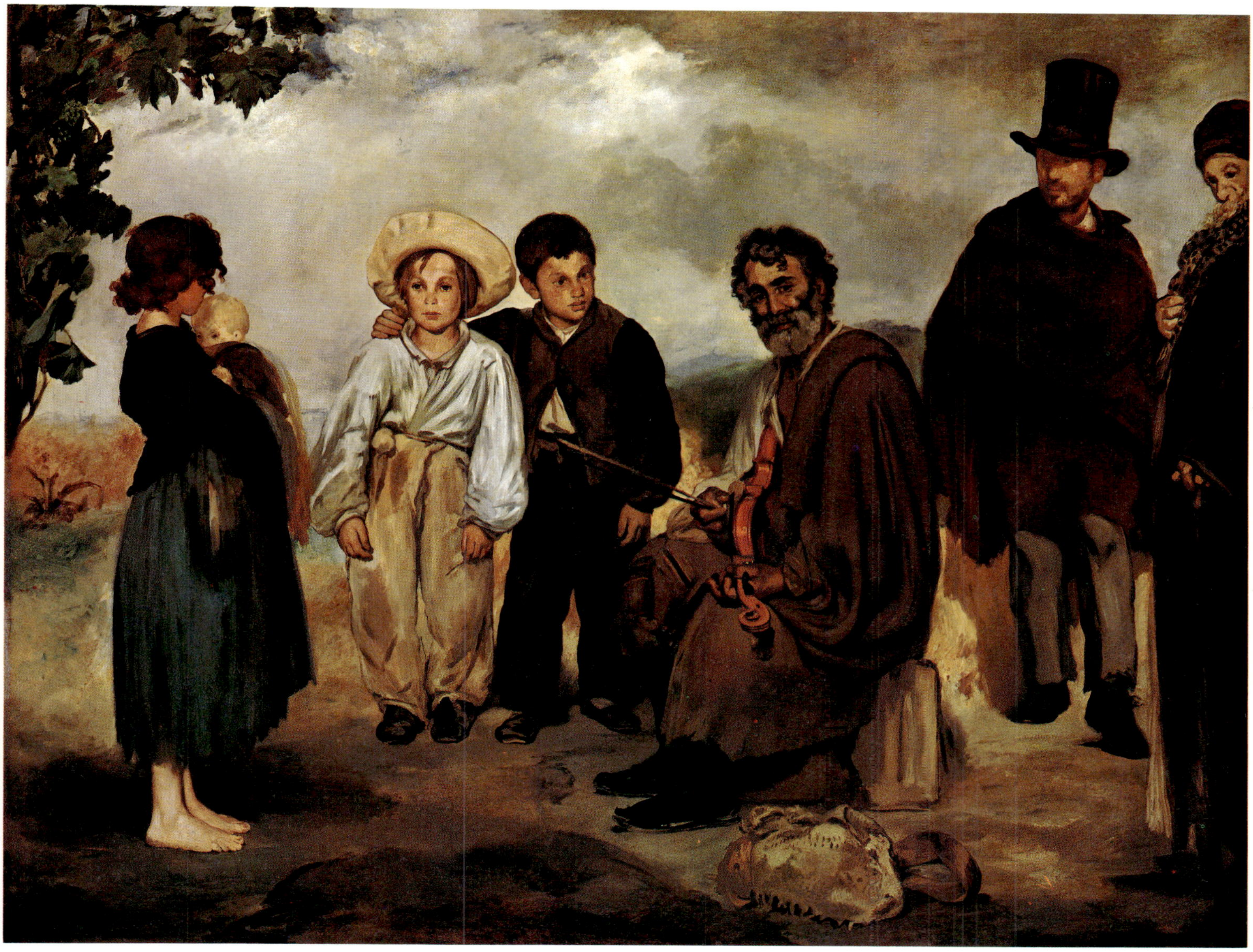

50. EDOUARD MANET. *The Old Musician*. 1862. Canvas, 73 3/4 × 98″. Chester Dale Collection

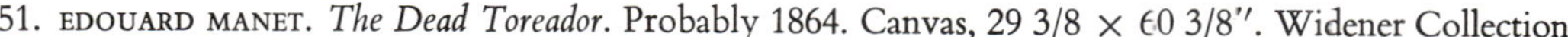

51. EDOUARD MANET. *The Dead Toreador*. Probably 1864. Canvas, 29 3/8 × 60 3/8″. Widener Collection

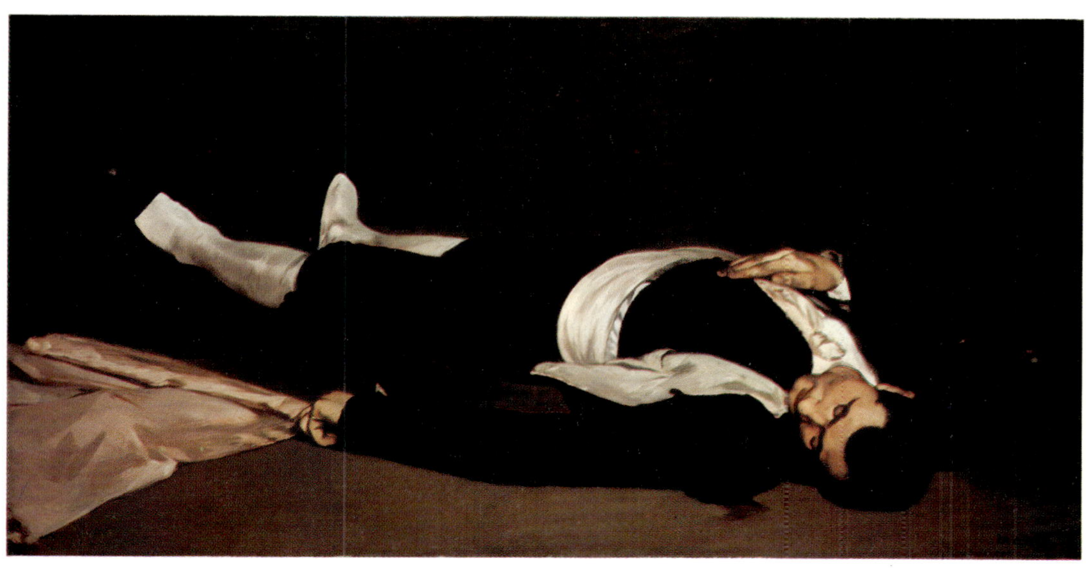

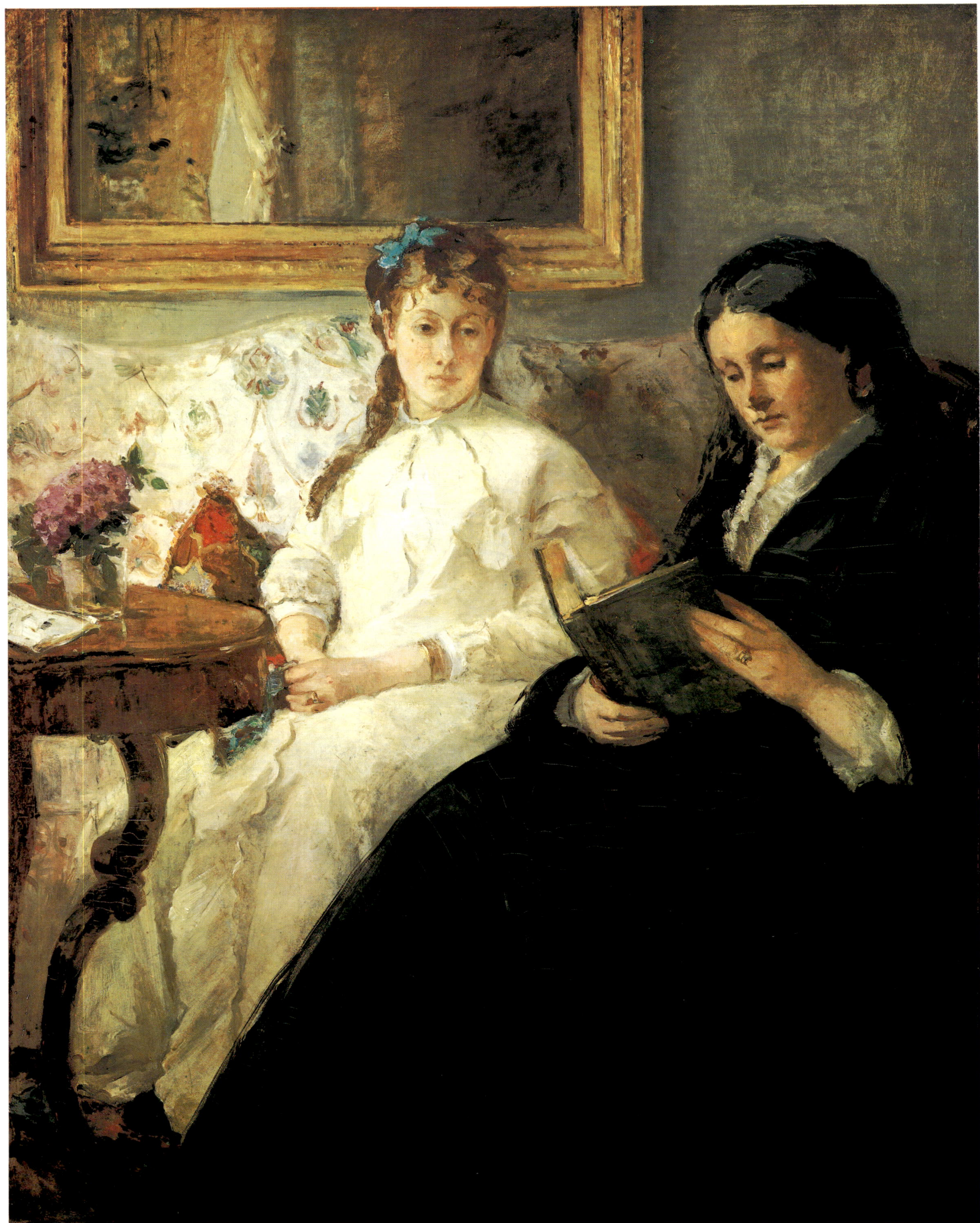

52. BERTHE MORISOT. *The Mother and Sister of the Artist.* 1869–70. Canvas, 17 1/8 × 18 1/4″. Chester Dale Collection

BERTHE MORISOT

The Mother and Sister of the Artist

Berthe Morisot was fascinated by the play of light on objects and delighted in painting them with freshness and delicacy. A critic of her time commented, "All her work is bathed in brightness, in azure, in sunlight. It is a woman's work, but it has strength, a freedom of touch, and an originality which we would hardly have expected."

In the late 1860s Morisot painted a number of family portraits, among which was the portrait of her mother and sister shown here.

On the gaily patterned floral sofa, resting against a colorful pillow accented with touches of red, is a young woman dressed in white. A blue bow is tied in her hair. She seems pensive and lost in thought. Her mother sits nearby, absorbed in a book. Mme Morisot is dressed in black, her somber costume relieved only by a bit of white at the neck and cuffs. The contrast between the two figures is striking. In front of the sofa is a mahogany table on which are some violets in a glass and a piece of paper. Above the sofa is a picture.

Berthe Morisot was not satisfied with this painting, and because she wanted to enter it in the Salon exhibition of 1870, she asked the artist Edouard Manet for his criticism. Manet praised the painting and offered a few suggestions. However, he did more; he picked up a brush and worked for four hours retouching the portrait of the mother, whose face has a sculptural quality unlike that of the daughter. Although greatly annoyed, Morisot could not undo Manet's handiwork—a messenger from the Salon stood waiting to take the picture. The painting, however, was well received. It serves as a unique example of the work of two artists, combining Morisot's delicacy and freshness of color and quivering light with Manet's darker, heavier painting.

Berthe Morisot

Berthe Morisot was born in 1841 in Bourges, France, but soon moved to Paris with her family. As a girl, she took private art lessons with her sister Edma, and later they studied open-air painting with Corot and traveled throughout the French countryside. One day while painting in the Louvre, Morisot met Edouard Manet, whose influence eventually caused her technique to become freer. Several years later, she married Manet's brother Eugène and exhibited at the first independent showing of the Impressionists. During the last years of her life, Berthe Morisot's work came to be greatly admired. She died in 1895.

53. AUGUSTE RENOIR. *A Girl with a Watering Can.* 1876. Canvas, 39 1/2 × 28 3/4″. Chester Dale Collection

PIERRE-AUGUSTE RENOIR

A Girl with a Watering Can

A Girl with a Watering Can is one of Renoir's most charming pictures. Not only are we enchanted by the delicate young girl standing in the middle of the garden path; we are attracted immediately by the luminous colors and shimmering paint used by the artist.

The rose-red of the saucy bow in the little girl's hair is repeated in the red of her lips and the reds and pinks of the flowers. The rich blue of her dress is echoed in the watering can, which is blended with the dress in such a way that it seems almost a part of it; in her eyes; in her shoes; and, here and there, in a flower. Intense shades of green give a lush quality to the grass and foliage. The garden path is warm with sunlight; these same golden tones light up the child's hair and face. She seems proud of herself, and stands confidently, as if aware of the artist painting her.

Painted when Renoir was deeply involved with the methods and aims of Impressionism and concerned with creating compositions filled with light, the picture is composed of small strokes of glowing colors. The pinks and reds, yellows, greens, and blues, heightened by touches of white, blend to create a lustrous, sparkling surface.

Pierre-Auguste Renoir

Born in Limoges, France, in 1841, Pierre-Auguste Renoir was apprenticed at thirteen to paint dishware for a porcelain manufacturer in Paris, where his family had come to live. Later he earned money decorating fans and began to study painting alongside such young artists as Monet and Bazille. Renoir became one of the most important members of the Impressionist group. Some of his favorite subjects were beautiful women, children, and gay outdoor genre scenes. During his last years his hands became so crippled with arthritis that he had to have his brush strapped to his hand in order to continue Painting. Nonetheless, Renoir's late pictures, like his more youthful works, bear witness to his often-expressed intention: "The earth as a paradise of the gods, that is what I want to paint." The artist died in 1919.

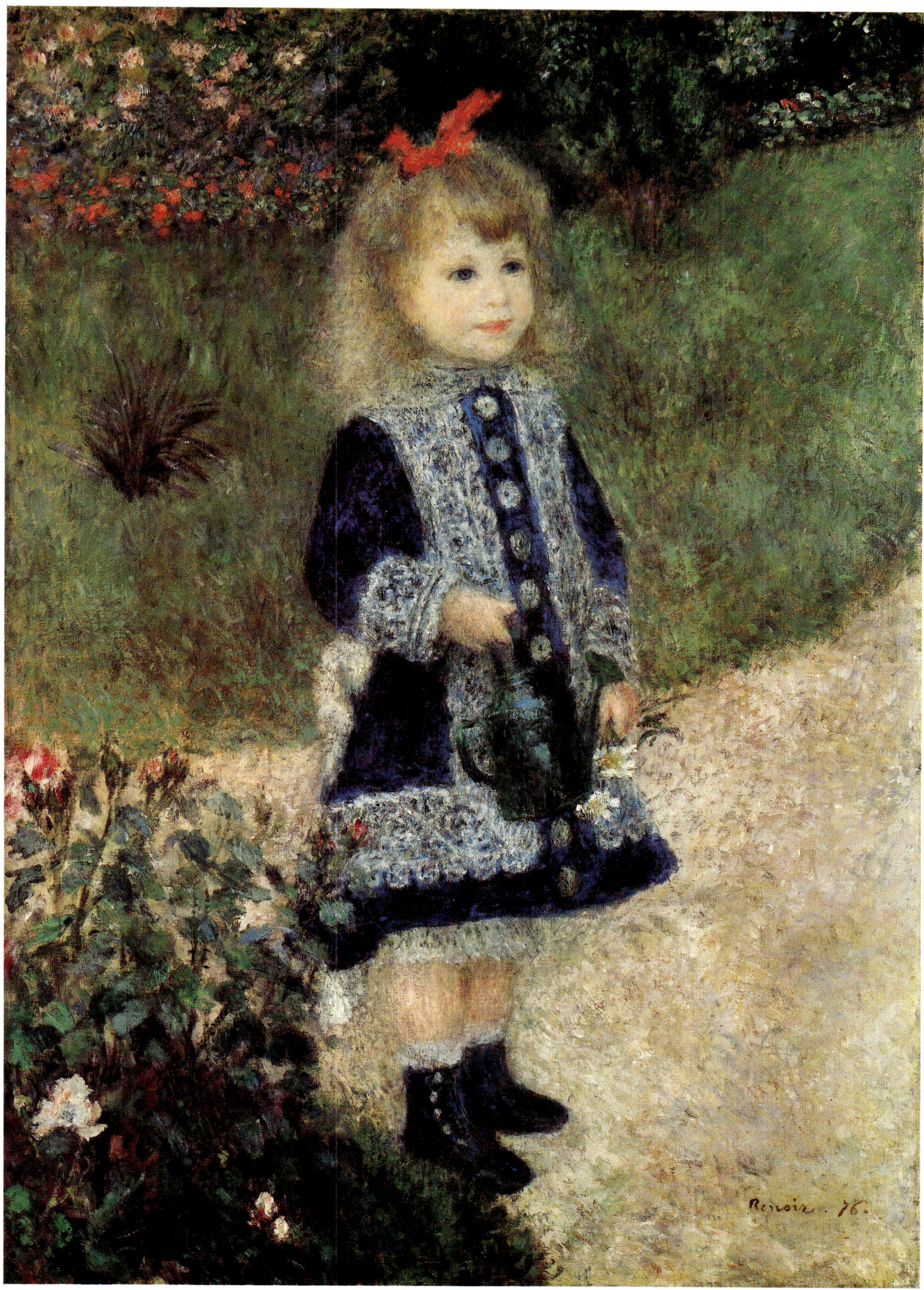
Renoir. 76.

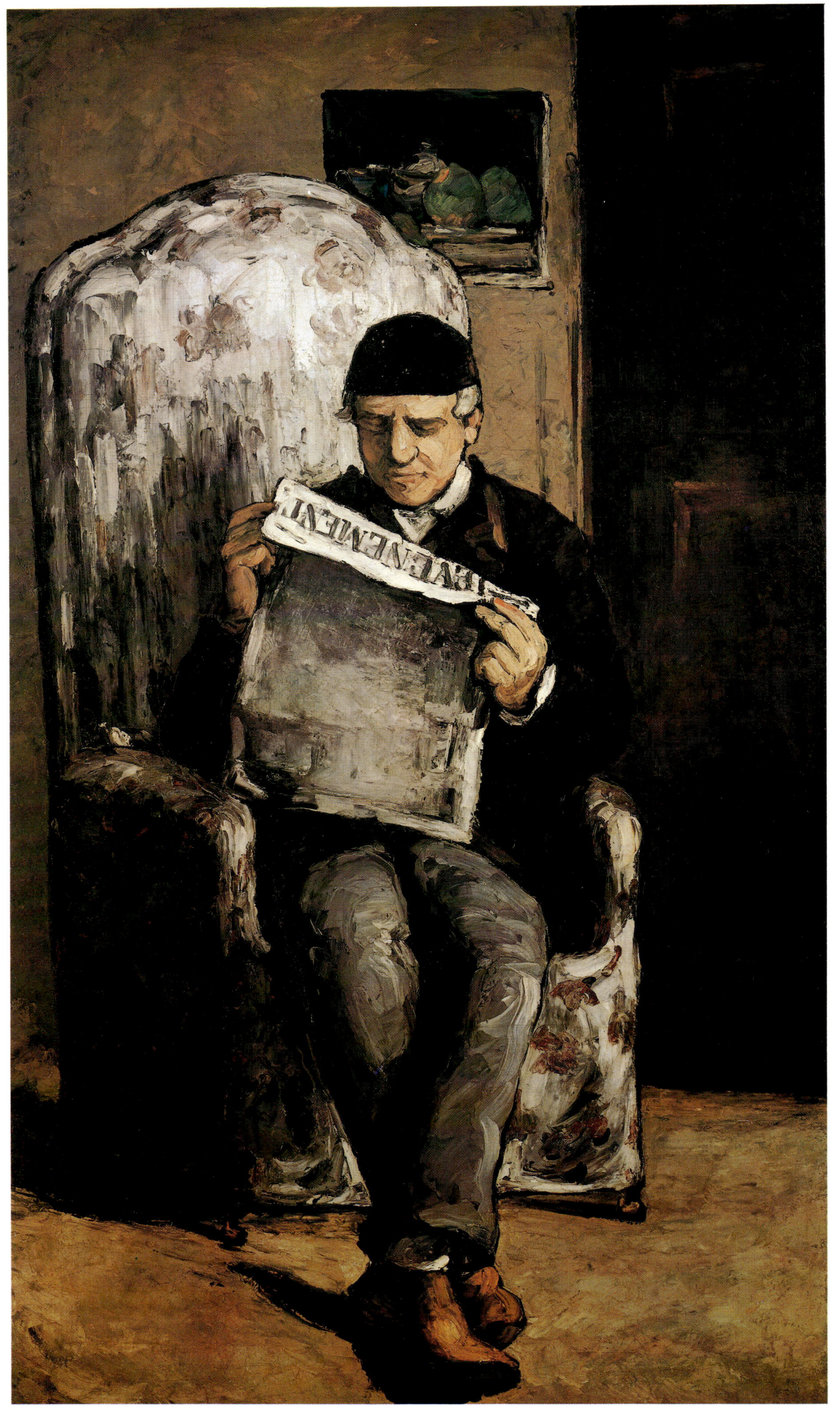

PAUL CÉZANNE

The Artist's Father

The Artist's Son, Paul

Strong ties existed between the middle-class businessman Louis-Auguste Cézanne and his moody artist son, Paul.

In this powerful, compassionate portrait of his father, Cézanne has made us feel these ties and has given us a sense of the authority and quiet dignity of the man.

The artist has depicted his father seated in a large, high-backed, gray-white armchair, reading a newspaper. The man's figure is drawn in such a way as to suggest solidity and permanence, and its angled position gives depth to the space. This illusion of depth is enhanced by the shadow on the floor under the man's foot and by the artist's early still life hanging on the wall behind the figure. Cézanne has deliberately changed the perspective by flattening out the back of the chair and tilting the floor forward.

Cézanne advised artists to look for "the cone, the sphere, and the cylinder" in nature. He searched for the basic form of each individual object before he began painting. This picture reflects the artist's interest in the shapes of things and in their relationship to each other. He contrasts the man's round head with the rectangles of the newspaper and chair, set off by heavy paint applied in ridges and lumps. The delicate balance between the controlled, simple order of the stable forms and the energetic application of paint gives the painting great force.

Cézanne's portrait of his son, Paul, shows a young man with a self-confident, level-headed, yet carefree air—qualities Cézanne himself lacked.

The background diamond shapes do not cover the entire canvas. Spots of canvas have been deliberately left bare in order to set off and brighten the design.

Against this background Cézanne has posed young Paul. A dark bowler hat perched at a jaunty angle sits on his solidly structured head. His large round dark eyes, arched brows, curved nose, and protruding lower lip are forcefully drawn, giving his face a sculptural quality. The blue single-breasted jacket is painted in such a way that it gives the feel of volume and weight to both the cloth and the boy's body. Cézanne has used the angles of the folds at the left elbow of the jacket and the notches of the collar to unify the shapes in the diamond background.

Cézanne added something new to the bright colors and small strokes of Impressionist painting by depicting his subjects in terms of planes, or facets, which he builds into unified forms which have weight, volume, and, for the viewer, eternal solidity.

◁ 54. PAUL CÉZANNE. *The Artist's Father.* 1866. Canvas, 78 1/8 × 47″. Collection of Mr. and Mrs. Paul Mellon

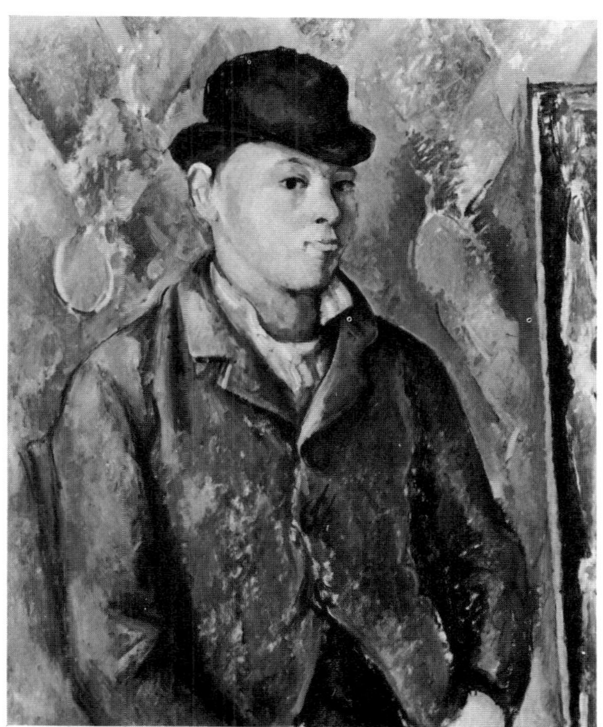

55. PAUL CÉZANNE. *The Artist's Son, Paul.* 1889/90. Canvas, 25 3/4 × 21 1/4″. Chester Dale Collection

Paul Cézanne

Paul Cézanne was born in the small town of Aix-en-Provence in southern France in 1839. Paul's father insisted that he attend law school and enter the family banking business, but when he proved a failure at both law and finance, he revolted and went to Paris to study painting. Cézanne's earliest pictures were inspired by painters like Delacroix, Courbet, and Velázquez, but he soon decided to seek a durable, timeless art which would make use of the innovations of his Impressionist friends. The result was a new kind of picture with recognizable subject matter but an underlying structure of colors and shapes so strong that it seemed almost abstract. Soon after he died in 1906, Cézanne was already being called the "Father of Modern Art."

VINCENT VAN GOGH

The Olive Orchard

In *The Olive Orchard* Vincent van Gogh has portrayed three women harvesting olives. The very old and gnarled trees are depicted in lilac-brown and yellow earth tones with silvery-green leaves. Similar muted shades are used in depicting the ground. In the center, the women, dressed in pale colors, carefully place the olives into baskets hung from the branches of the trees. The dark outlines of their dresses make them stand out. Quick brushstrokes in pale pinks and greenish yellows delineate the sky. There is a swirling rhythm to the scene created by the short, sharp, wavy brushstrokes in the foreground and in the trees, which, with their writhing trunks, seem alive with motion. The soft, delicate tones create an impression of nostalgic sadness.

Before painting this picture, Van Gogh wrote enthusiastically to his brother, "Oh my dear Theo, if you could see the olive groves just now! . . . The rustle of an olive grove has something very secret in it, and immensely old." In a later letter to Theo, Van Gogh described the painting as "a canvas done from memory after the study of the same size made on the spot, because I want something far away, like a vague memory softened by time."

All these impressions are reflected in the painting, a hymn in subtle tones to the mysterious forces of nature, with which Van Gogh was so much in tune.

Vincent van Gogh

Vincent van Gogh was born in the Dutch village of Groot Zundert in 1853. As a young man he worked for his uncle, an art dealer, until his sympathy for the poor peasants and miners of Holland prompted him to take up missionary work. He soon realized he was unsuited to this vocation, but he began to express his feelings for these deprived people by doing paintings of them. During this period he lived in poverty himself, and he became increasingly ill and dispirited. He finally decided to join his brother Theo in Paris. There he became acquainted with the bright paintings of the Impressionists, and his own somber paintings took on a new vitality. In 1888, Van Gogh moved to Arles, in the south of France, where he produced many of his finest paintings. His mental instability grew more serious, however, and he finally died by his own hand in 1890.

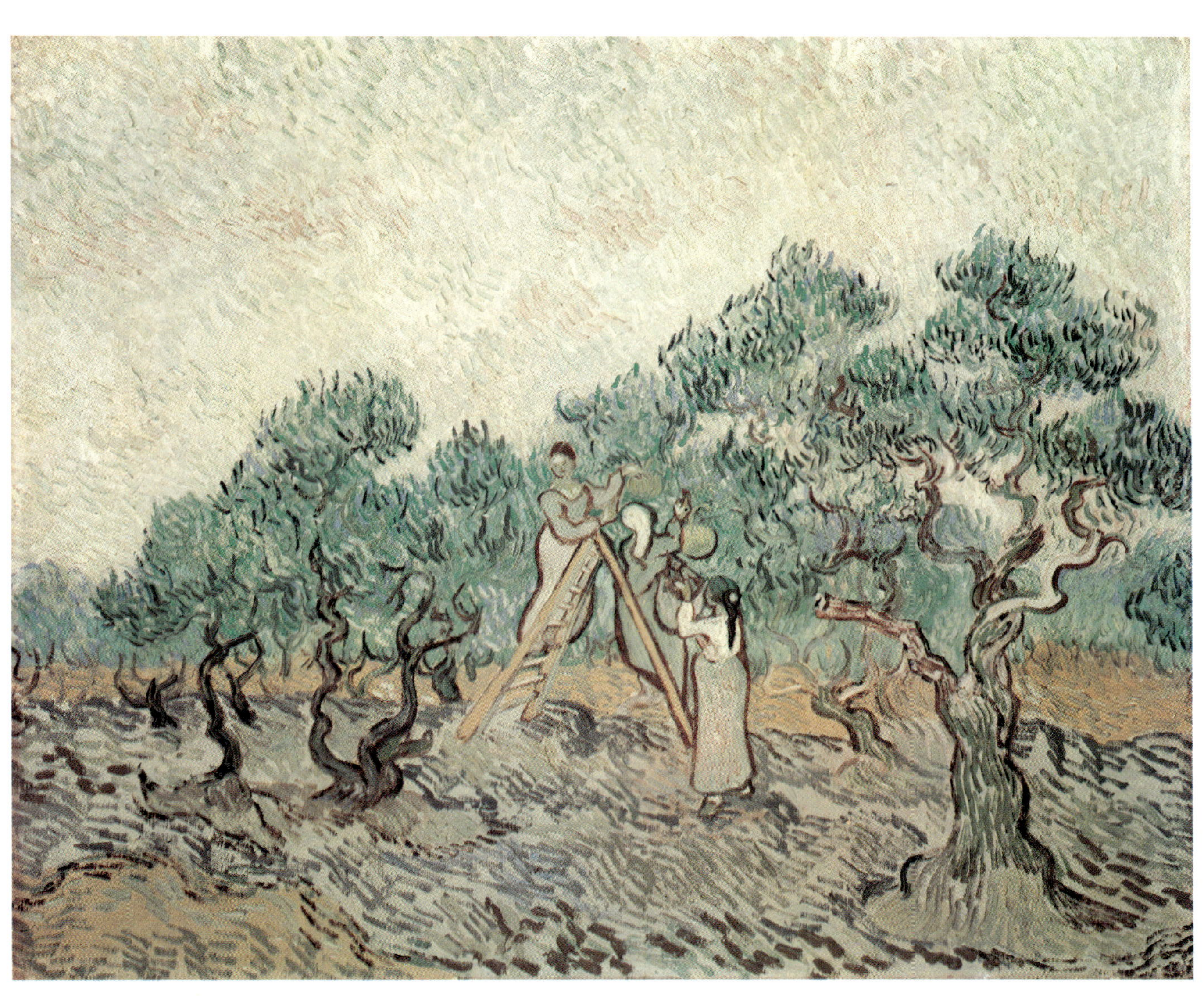

56. VINCENT VAN GOGH. *The Olive Orchard.* 1889. Canvas, 28 3/8 × 36 1/4″. Chester Dale Collection

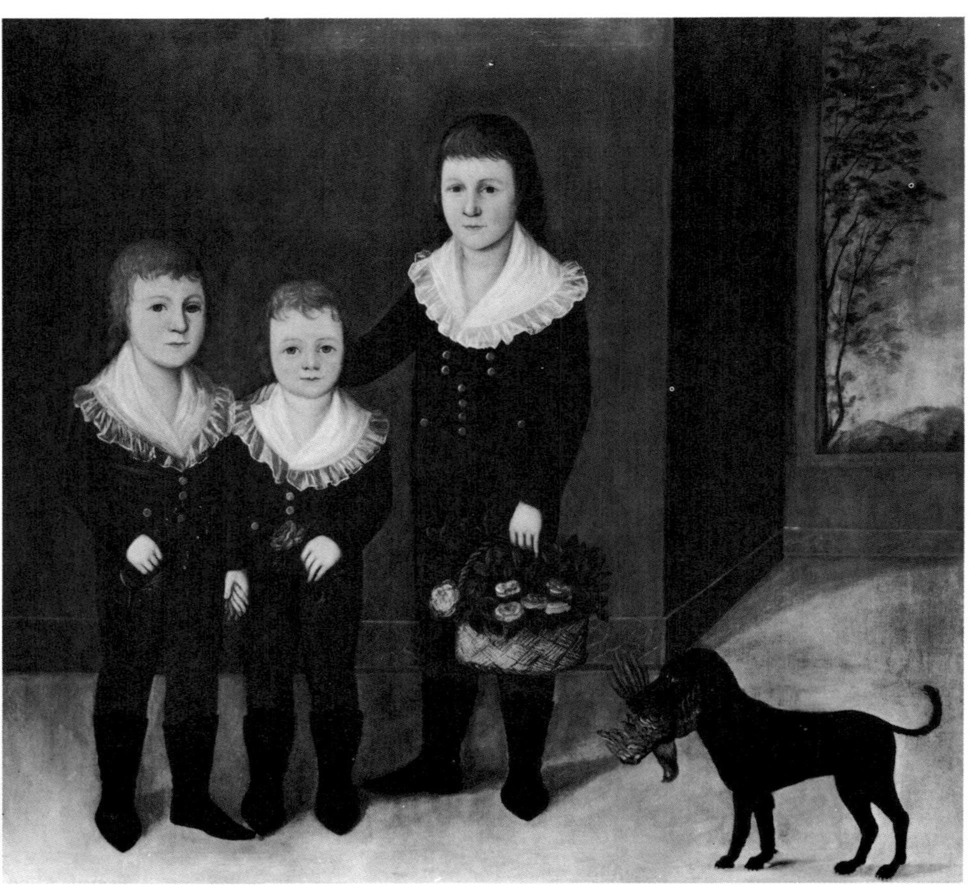

57. JOSHUA JOHNSTON. *The Westwood Children.* c. 1807. Canvas, 41 1/8 × 46″. Gift of Edgar William and Bernice Chrysler Garbisch

JOSHUA JOHNSTON

The Westwood Children

The Westwood children, members of a well-to-do family of Baltimore, Maryland, were painted by Joshua Johnston, considered by many the first important black artist in the United States. He has shown the boys in a simple setting, standing before a wall. Behind them, seen through a large window, is a tall leafy tree and distant rolling hills.

Henry, on the left, holds the hand of his brother George Washington Westwood, while John, the eldest, rests his hand lightly on the shoulder of his youngest brother. The boys, who are dressed alike in jumper suits with large ruffled collars, bear a marked resemblance to one another. They have rather serious expressions on their faces. Each boy holds something in his hand—a sprig of cherries, a rose, a basket of flowers. The silhouetted form of the dog holding a bird in its mouth serves to balance the three figures and supplies an appealing decorative touch.

Johnston had not mastered the techniques for representing figures and space in a naturalistic manner. The boys and dog are composed of flatly painted, unmodulated shapes; the proportions, too, might strike us as a bit strange, for the heads are too large for the bodies. Despite these technical limitations, the crisp outlines and charming details make this picture particularly delightful.

Joshua Johnston

Joshua Johnston, a portrait painter active in Baltimore from 1796 to 1824, was born about 1770 and is believed to have been a slave. What little we know about Johnston was discovered through careful detective work in the city records, and paintings could not be firmly attributed to him until recently, when a signed portrait finally came to light. Today Joshua Johnston's portraits are prized possessions of the museums lucky enough to own them.

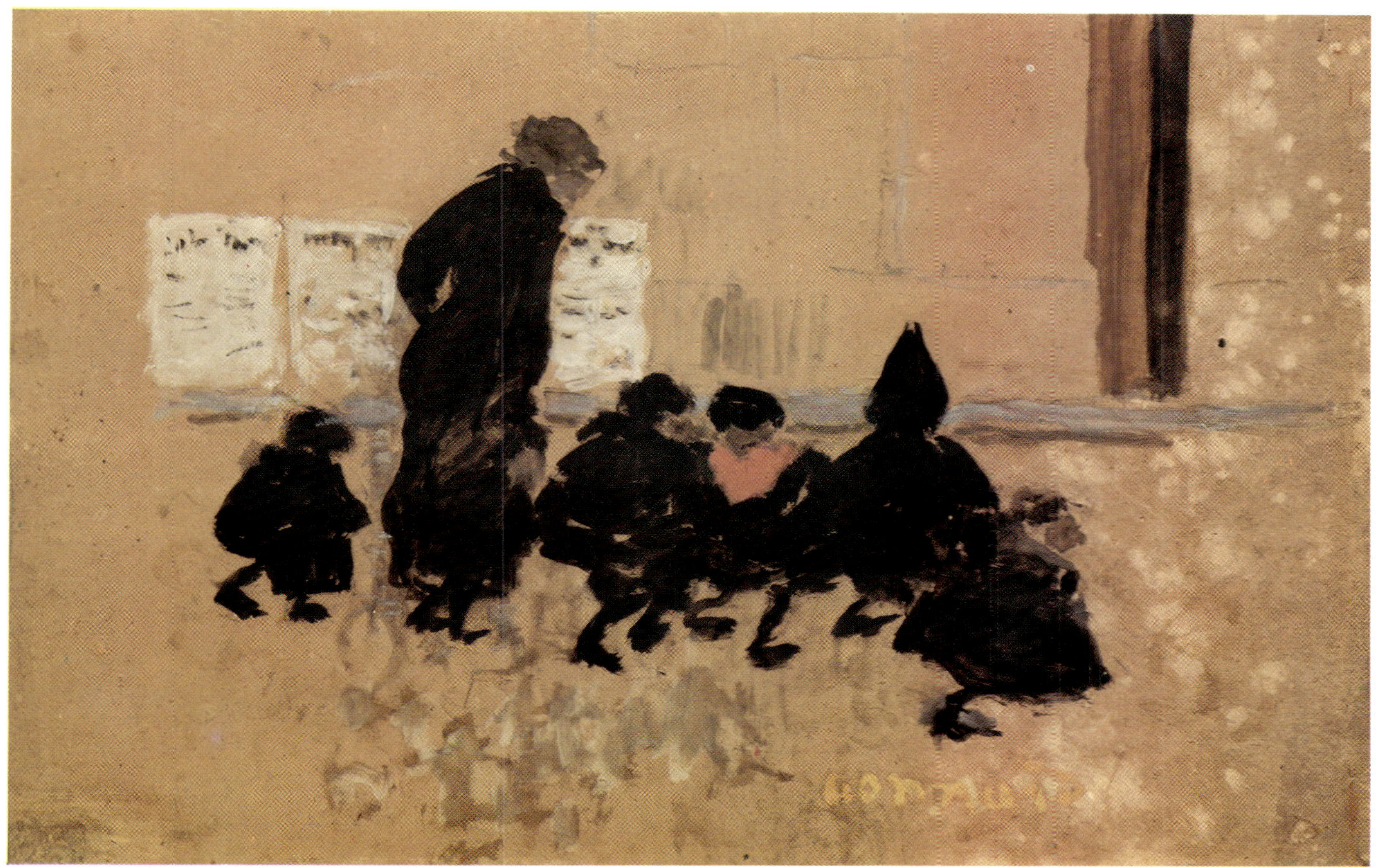

58. PIERRE BONNARD. *Children Leaving School.* c. 1895. Cardboard on wood, 11 3/8 × 17 3/8″. Ailsa Mellon Bruce Collection

PIERRE BONNARD

Children Leaving School

What a sense of delight we feel when viewing this painting by Bonnard! With a few roughly drawn shapes, quickly sketched lines, and touches of color we are given a complete picture.

Bunched together in the center of the painting are blotchy silhouettes representing children returning home from school. At the head of the group is a child wearing a cape with a pointed hood. The others wear small caps and capes. The red accent on one of the figures represents a scarf; it serves as the focal point of the painting. The figure of the woman, perhaps a governess, in long coat and bonnet, is awkwardly bent forward as she moves with the group of children. Behind her, lagging a bit, is the smallest child, tugging his schoolbag. Two heavy lines on the right suggest the end of the building and the intersection of another street. The blue-gray lines indicate the bottom of the building. Bonnard has used touches of pale color to delineate the ground.

The white shapes with black markings on the wall in the background represent advertising posters.

In this picture Bonnard has captured with tenderness and sensitivity an experience with which we can all identify. He has also created a whimsical, decorative pattern which is enormously appealing.

Pierre Bonnard

Born in a small town near Paris in 1867, Pierre Bonnard showed no interest in art until he completed school and began to work as a government clerk. It was then that he began to take evening art classes and met Edouard Vuillard, who would be a central figure in the art movements succeeding Impressionism. Bonnard and Vuillard shared a preference for intimate, everyday scenes, painted in a flat, loosely brushed manner. Both were great admirers of Gauguin and Japanese prints. Bonnard exhibited his work in important group shows and later alone, with great success. In 1925, he married his model of many years and moved to a small house on the Riviera. He died there in 1947.

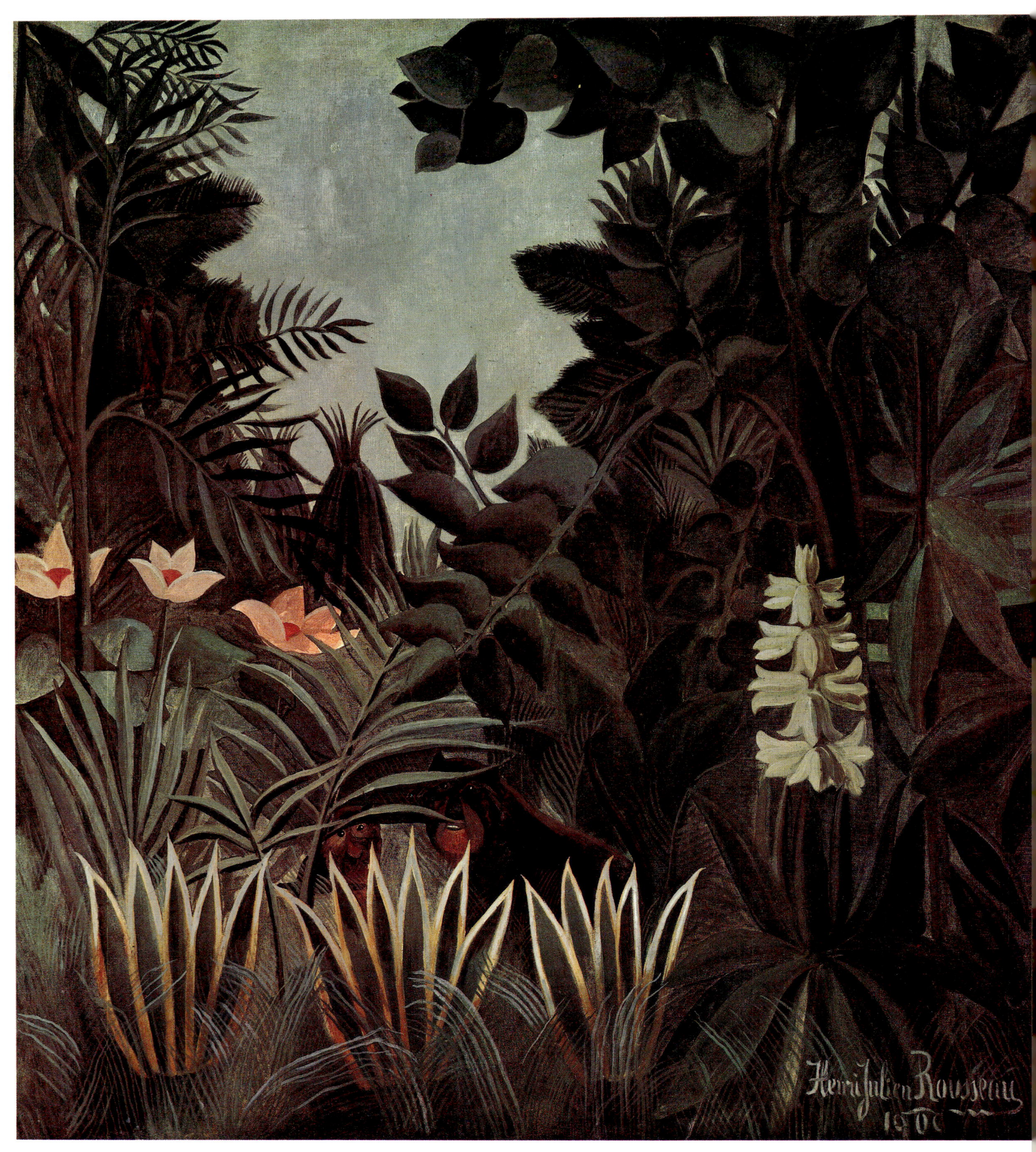

59. HENRI ROUSSEAU. *The Equatorial Jungle*. 1909. Canvas, 55 1/4 × 51″. Chester Dale Collection

HENRI ROUSSEAU

The Equatorial Jungle

UNKNOWN PAINTER

The Dog

In *The Equatorial Jungle*, one of a number of jungle pictures he painted, Henri Rousseau has created a wild and menacing exotic scene. The oversized, thick, spiky, interwoven gray-green leaves and grasses make us feel the lush dampness of the jungle. We can almost hear the eerie sounds and sense the hidden life in the dense foliage, which parts to show an unworldly gray-green sky.

From the rhythmically repeated leaf shapes, gigantic stalks of hyacinth and mother-in-law's tongue, and exotic pale flowers, two dark brown beady-eyed creatures peer out. In appearance they seem a comic cross between a baboon and an orangutan. Above them a strange bird with brown breast and head and black dots for eyes perches on a curving branch. Rows of plants across the front, with tall upward stalks at the sides, create a bold arrangement of simple rhythmic flat designs and patterns with subtly varied hues.

This threatening jungle comes alive in a strange and magical way. It is said that while working on his jungle paintings Rousseau would often become so frightened that he would run to open a window to break the picture's spell on him.

The untutored American folk-art painter who created *The Dog* displays the same instinctive understanding of harmony, color, balance, design, and structure that we see in Rousseau's work.

The dog is portrayed against a streaked sky, in front of a very low brick wall, his bushy tail curled over his back. He has clear small eyes and his mouth is open, showing two fanglike teeth. He seems to be smiling. The animal is placed behind an edge of walk upon which are set at precise distances seven tiny pots of plants. The artist has emphasized the dog's importance by making him gigantic in comparison with the other objects in the painting.

On top of the raised cornices of the wall, forming a decorative pattern on either side of the painting, are cement planters containing a variety of flowers. Three arrangements of vines are spaced at intervals along the wall. These are set off by tiny flowers and pebbles arranged in a semicircle. Like the dog, the plants and flowers are depicted with attention to their decorative appeal rather than to their natural appearance. We have no sense of their real texture or size but instead are charmed by the lively contrasts of curving and straight lines.

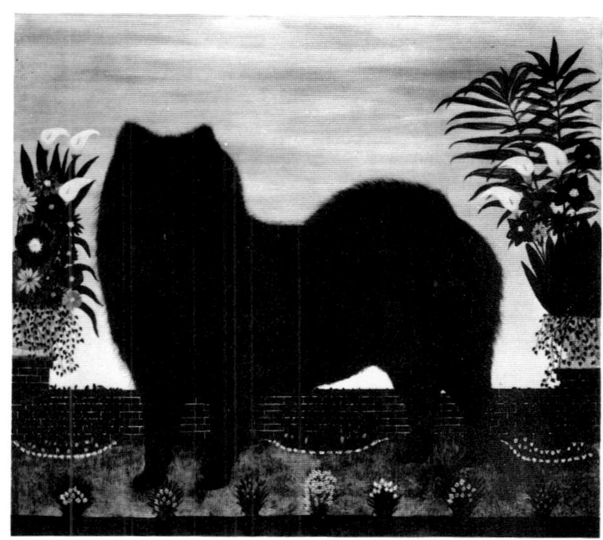

60. UNKNOWN PAINTER. *The Dog*. c. 1860. Canvas, 35 1/4 × 41 1/2". Collection of Edgar William and Bernice Chrysler Garbisch

Henri Rousseau

The son of a tinsmith, Henri Rousseau was born in Laval, France, in 1844. Except for two brief periods in the army, he earned his living as a *douanier*, or French customs official, until 1885. For this reason he was nicknamed "Le Douanier" Rousseau by the art world. Rousseau began painting as a hobby, and never received any formal training. His technique was thus extremely naive; for example, he measured people like a tailor before painting their portraits. But Rousseau's paintings were so beautiful and magical that artists like Gauguin, Van Gogh, and Picasso immediately appreciated their value and invited him to exhibit with them. Rousseau died in 1910.

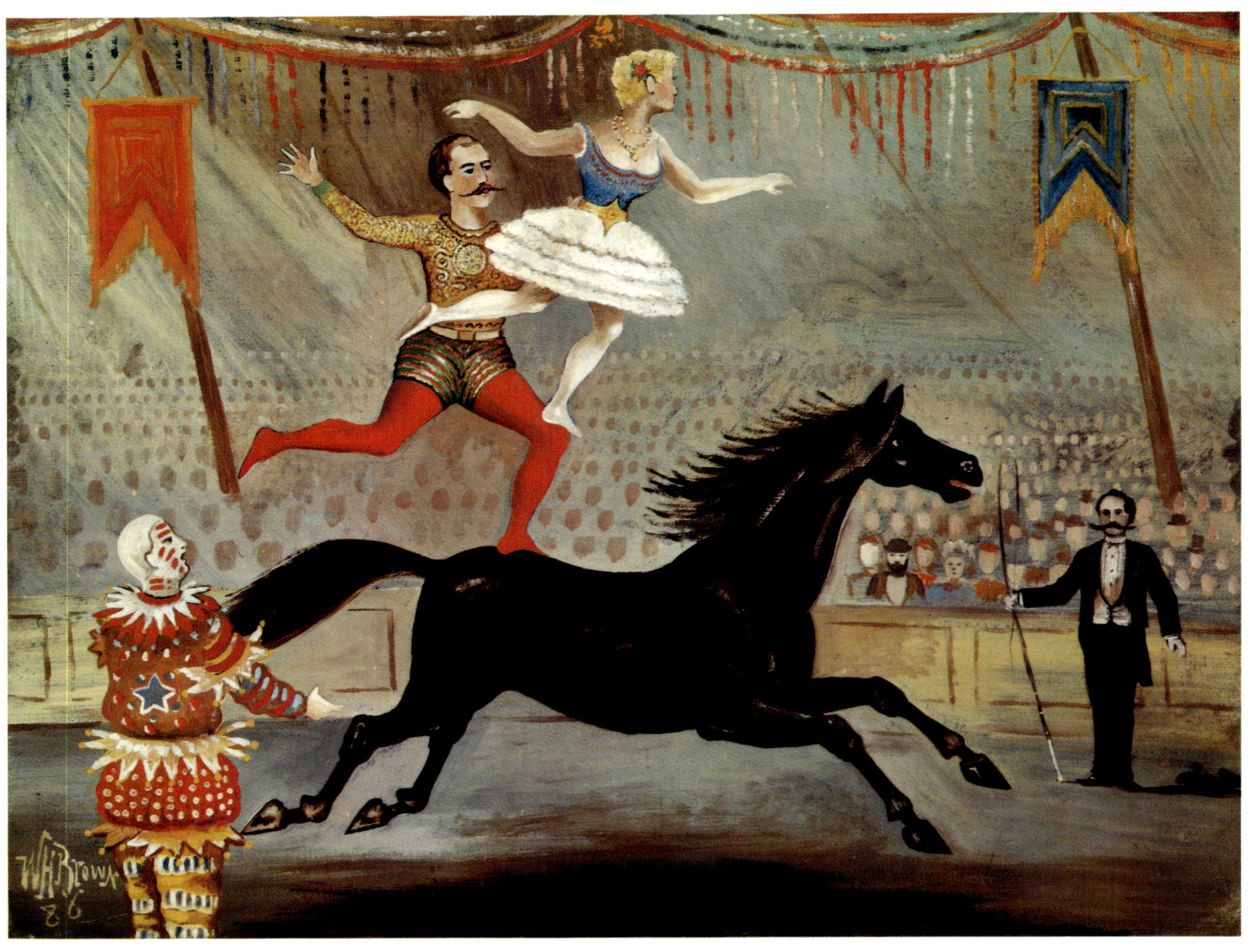

61. W. H. BROWN. *Bareback Riders*. 1886. Cardboard, 18 1/2 × 24 1/2″. Gift of Edgar William and Bernice Chrysler Garbisch

W. H. BROWN

Bareback Riders

UNKNOWN MEDIEVAL SCULPTOR

Aquamanile in the Form of a Horseman

In *Bareback Riders* the American folk artist W. H. Brown has caught all the glamour, fascination, and excitement of the circus. Brown must have laughed at the clown's pranks, admired the ringmaster's dignity, and marveled at the agility of the bareback acrobats. Then he returned home to paint the scenes on a piece of cardboard.

In the foreground of the painting is a galloping black horse. A male acrobat stands on the horse's back; one leg is bent at the knee to support the gracefully poised figure of his daring partner. The arms of both performers are extended to emphasize the sense of risk of the act. Their colorful costumes add to the gaiety of the scene.

On the left of the picture is a clown, his chalk-white face daubed with red paint. On the other side, acting as a balance for the clown, is the ringmaster, who wears a formal black suit and holds a long whip. His demeanor and costume reflect his important position in the circus.

Brown's way of painting reflects his lack of formal schooling in art. He presented his subjects in a simple and direct manner, outlining the forms so that each figure is a flat shape, like a paper cutout. The rigid, frozen appearance of the figures results from Brown's inability to depict people as they actually look when they are in motion. The clown's head is abruptly twisted into profile; such side views of the face are relatively easy to draw. The artist did not know how to realistically render the lighting in the arena, but he did suggest the haze in a circus tent that partially obscures the audience in the stands.

Brown conveys the gaiety and glitter of the circus through the use of pure red, blue, and yellow without any subtle mixtures of color. He creates a feeling of movement by placing the clown, the man, and the woman on an imaginary diagonal line leading upward and by depicting the horse with mane flying and all four feet off the ground. The bareback riders seem to float in the air. By showing the partners in similar poses—even their pointed toes are at the same angles—the painter has caught the coordinated rhythm so necessary in acrobatic feats and has brought out the grace and skill of a thrilling performance.

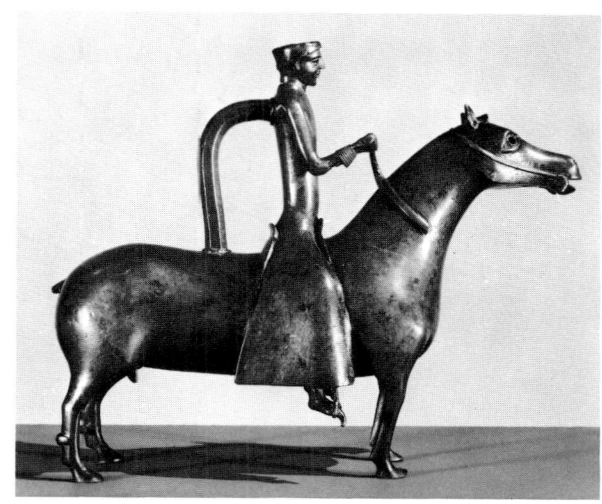

62. UNKNOWN SCULPTOR. *Aquamanile in the Form of a Horseman.* XIII century. Bronze, 11 1/4 × 14 × 6″. Widener Collection

The elaborate bronze statuette shown here is actually a ceremonial water pitcher. The word *aquamanile* combines two ancient Latin words: *aqua*, meaning "water," and *manare*, a verb meaning "to flow." Thus an aquamanile is a jug from which water is poured.

The unknown sculptor who cast this aquamanile probably lived in England or Germany during the thirteenth century. First, this skilled craftsman modeled his statuette in clay or wax. Then he made a mold from the statuette and discarded the original. After this he melted bronze in a furnace and poured the liquid metal into the hollow mold. When the metal cooled and hardened, the sculptor broke the mold, revealing the aquamanile. Finally, he engraved certain details, such as the cuffs on the gloves, and polished the metal to a high gloss.

The rider of this container-horse wears a long coat that hides his legs. His pointed-toe boots show in the stirrups, and he wears cuffed gloves to protect his hands from getting chaffed by the reins. His left arm is broken off and lost.

The sculptor showed great ingenuity in incorporating the working parts of the water pitcher into his design. Thus, the top of the horseman's hat serves as an opening through which the container can be filled, the mouth of the horse forms the spout, and a stylized version of the rider's cape forms the handle of this utilitarian but aesthetically unified object. To make sure that the jug would not tip over, the artist kept the horse's hooves firmly on the ground, and the rider sits with equal solidity on the animal's sturdy hollow back.

EDWARD HICKS

The Cornell Farm

PAUL GAUGUIN

Haystacks in Brittany

Edward Hicks

Edward Hicks, born in Pennsylvania in 1780, was a coach-painter by profession and painted pictures for pleasure in his spare time. He was also a Quaker preacher, and many of his paintings reflect his great religious faith. Today, his many versions of *The Peaceable Kingdom*, showing wild and tame animals living happily together, are especially admired for their beauty and simplicity. Hicks died in 1849.

Paul Gauguin

Born in Paris in 1848, Paul Gauguin was first a sailor and later a stockbroker before he began to paint. He showed his work with the leading Impressionist painters, but was thirty-five before he gave up the business world entirely. Then, becoming convinced that modern civilization was corrupt and unfeeling, he began to seek out the simple folk whom he felt were still innocent and good. He went first to the French province of Brittany, later to Panama and Martinique, and finally to Tahiti and the Marquesas Islands, where, except for one brief visit to Paris, he lived, painted, and wrote until his death in 1903.

Edward Hicks had no academic artistic background. He was commissioned to do *The Cornell Farm* by the proud farmer, who wanted a visual remembrance of his animals who had won awards at the local agricultural fair.

The artist painted with strict attention to detail. Across the front of the painting, posed as though they were still up for judging, are white, brown-and-white, and black-and-white cows; a red bull; white, gray, and red horses; several brown ponies; white sheep; and grayish-black pigs. A number of disproportionately small figures are scattered behind the livestock. One of the central figures, probably James Cornell, wearing a top hat and gray belted coat, appears to be pointing at the livestock.

In contrast to the rounded forms of the animals are the geometric patterns of the well-groomed fields. The trees of the orchard, which are almost bare, are set in neat rows. The blurred outlines of rolling hills can be seen in the distance. Behind a picket fence is a large farmhouse with a red door and a group of red-roofed farm buildings. This group is balanced by the barn on the left.

Hicks has given a sense of depth to the painting by softly blending the colors and by blurring the forms in the background landscape. The simple rounded shapes of the animals, the inconsistent sizes of the figures in space, and the prominently outlined, brightly patterned forms create a lively, childlike effect.

In 1890, more than forty years after Edward Hicks had created his primitive painting of the Cornell farm, Paul Gauguin painted *Haystacks in Brittany*. It is interesting to compare the naive and detailed subject matter in Hicks's painting with the abstract forms in Gauguin's work.

The vividly colored shapes in *Haystacks in Brittany* form decorative patterns which seem to flow into each other with an almost musical rhythm. In the foreground of the painting are three black-and-cream cows and a milkmaid whose clothing is in the same colors. The pattern of the maid's dark sweater and white hat is repeated in the dark foliage sprinkled with white flowers. She acts as a pivot in this painting, linking the colors and patterns of the foreground with the fields which stretch beyond her. To the left, stripes of orange earth and green plants complement the colors of the design in the lower-right corner of the canvas. There, mingled with the shapes which form the cow's body, are odd shapes of a pumpkin color, possibly burlap sacks, set against a yellow-green haystack. Sitting like loaves of bread on the soft green grassy areas and tilled fields are rust-brown rectangular haystacks with rounded corners. In the distance, nestled in groves of lush green trees, are rust-colored shapes indicating the thatched roofs of houses. In contrast to the rich colors of the foreground, subdued tones create an overcast gray-blue cloudy sky, diffused with sunlight. This helps to convey the feeling of early autumn.

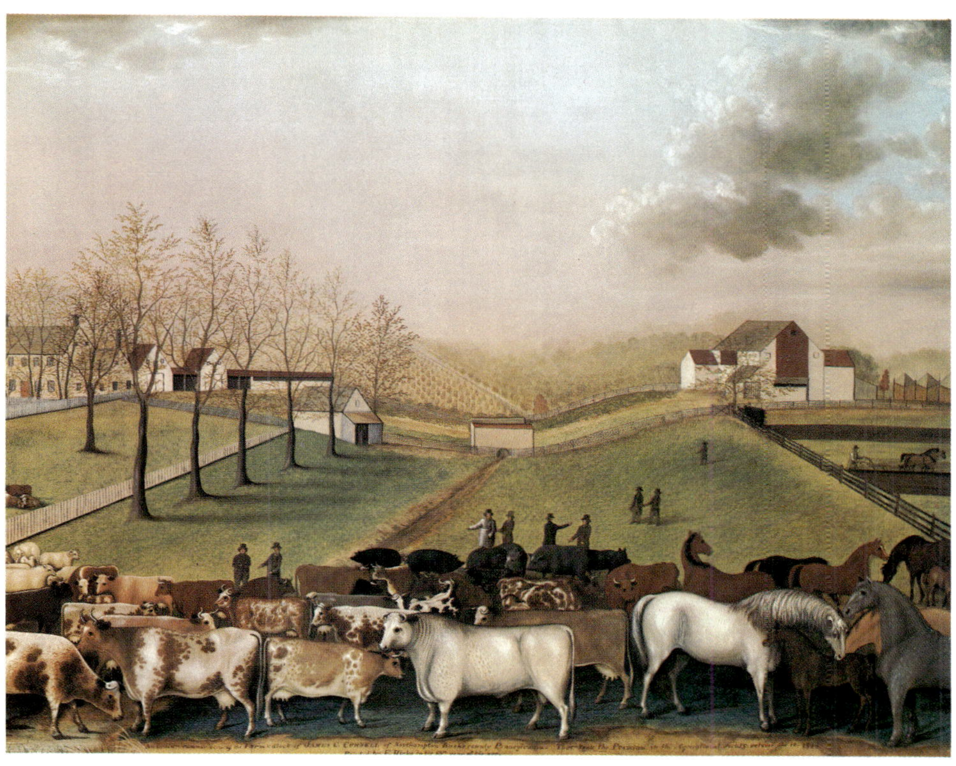

63. EDWARD HICKS. *The Cornell Farm.* 1848. Canvas, 36 3/4 × 49″. Gift of Edgar William and Bernice Chrysler Garbisch

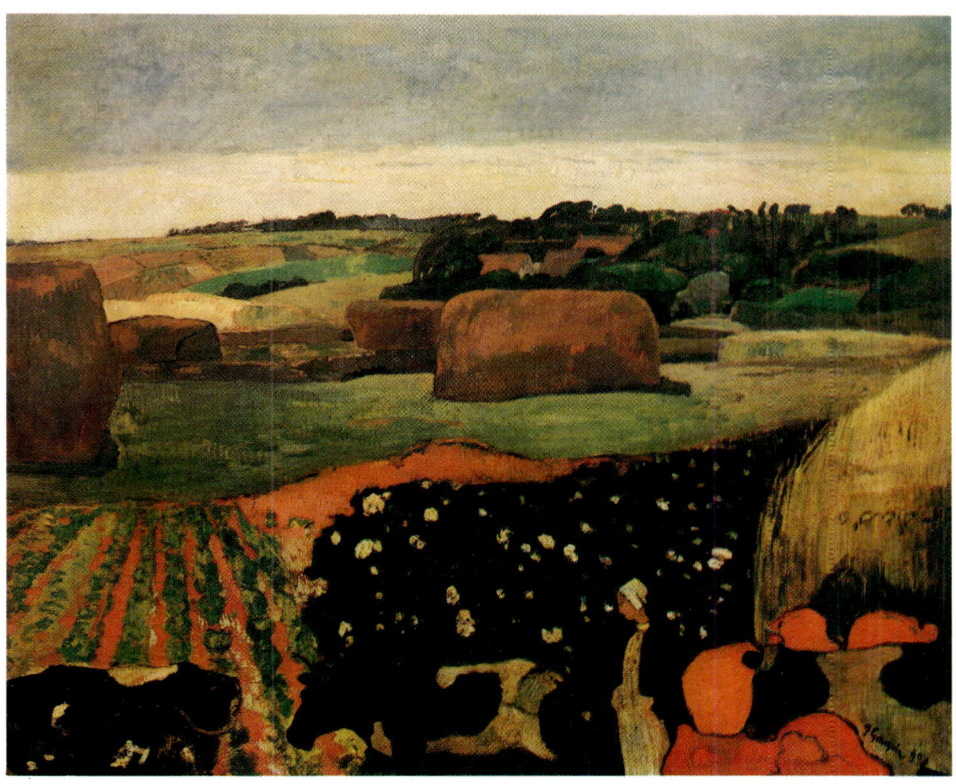

64. PAUL GAUGUIN. *Haystacks in Brittany.* 1890. Canvas, 29 1/4 × 36 7/8″. Gift of the W. Averell Harriman Foundation in memory of Marie N. Harriman

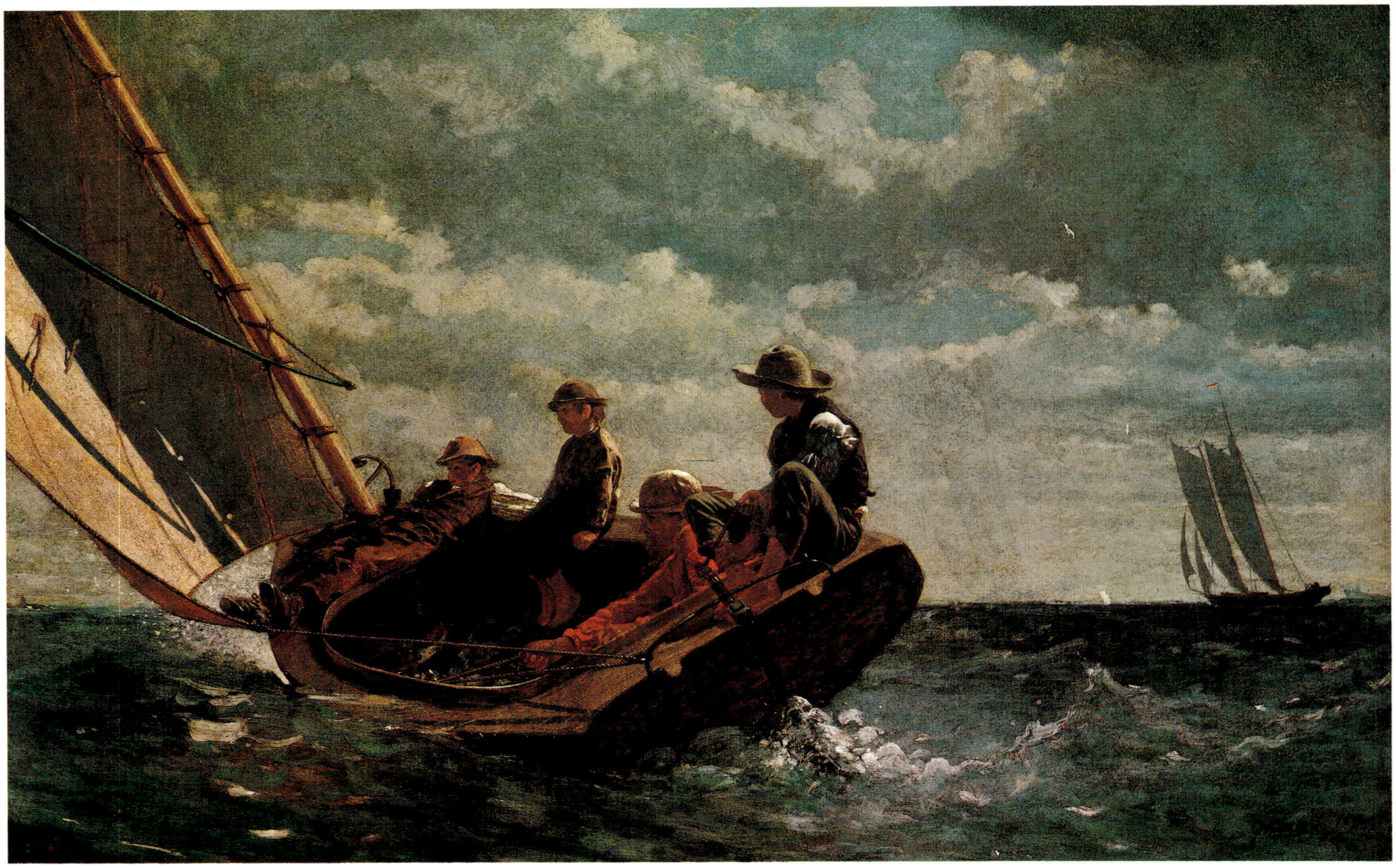

65. WINSLOW HOMER. *Breezing Up*. 1873–76. Canvas, 24 1/8 × 38 1/8″. Gift of the W. L. and May T. Mellon Foundation

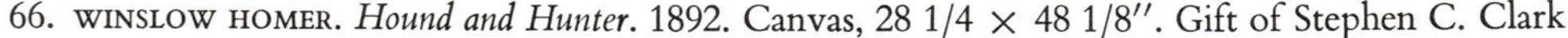

66. WINSLOW HOMER. *Hound and Hunter*. 1892. Canvas, 28 1/4 × 48 1/8″. Gift of Stephen C. Clark

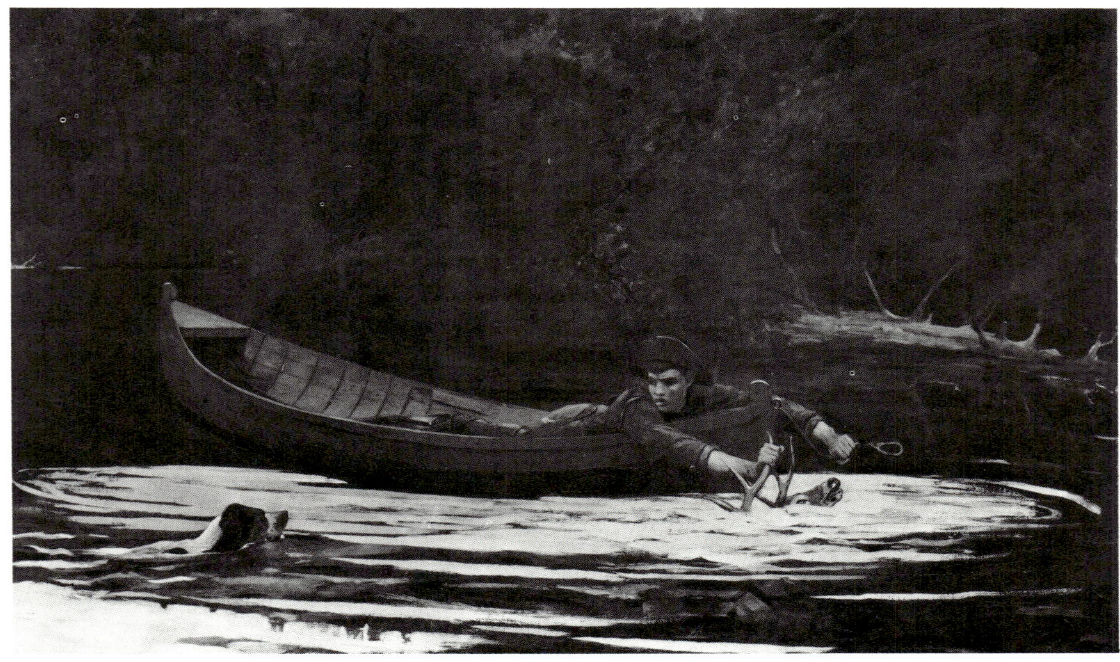

WINSLOW HOMER

Breezing Up

Hound and Hunter

A former illustrator, Winslow Homer painted in a careful, clear, accurately detailed, and convincing manner. Homer worked on *Breezing Up* at intervals over a period of three years. It was the result of intense study, and it grew out of two earlier studies of the scene, a watercolor and a small oil painting.

Sun-bronzed boys like these, in their weather-beaten clothes, were a common sight in New England in Homer's time, as were fishermen like the one in the red jacket, shown crouching as he holds the mainsheet. In the rising wind the boys have positioned themselves to counterbalance the tilt of the boat as it speeds along in a choppy sea. The lad stretched full length by the mast seems oblivious to the spray of the bow waves; the boy beside him, silhouetted against the sky, holds on to the coaming. The light, which highlights the figures of the sailors, also illuminates the scales of the fish in the bottom of the boat. The picture gives us a sense of the pleasure and independence of sailing.

Winslow Homer spent many vacations in the Adirondack Mountains in New York State and was familiar with the practices of hunters of the area. Dogs were used to drive deer into the water, where the deer became helpless and drowned. Homer may have reacted to this unsportsmanlike method of hunting by exposing it in *Hound and Hunter*, for he loved nature and its creatures. He has depicted the young hunter lying down in the stern of the boat, grasping the buck's antlers and at the same time keeping an anxious eye on the dog.

The head of the deer is almost completely submerged; the small portion of the eye we can see gives us an acute sense of the animal's panic. The artist has added the butt of a gun on the seat of the weathered boat. Fall foliage and a log along the shoreline form the background. There is a sense of force in the hunter's outstretched strong hand which grasps the deer's antlers. The dark waters are churned white, forming an ellipse that binds together the main pictorial elements of the action—the boat, the hunter, the hound, and the deer.

Winslow Homer

Winslow Homer, a descendant of an English sea captain and son of a talented watercolorist, was born in Boston in 1836. He was apprenticed to a lithographer, and soon earned his living as a free-lance magazine illustrator. Homer's first oil paintings came out of his experience as a correspondent during the Civil War. Later he concentrated on homely subjects typical of the American scene, such as children at play, cotton-pickers in the South, soldiers, fishermen, or bathers on a beach. He worked with equal zest in pen and ink, watercolor, and oil. Homer died in Prout's Neck, Maine, in 1910.

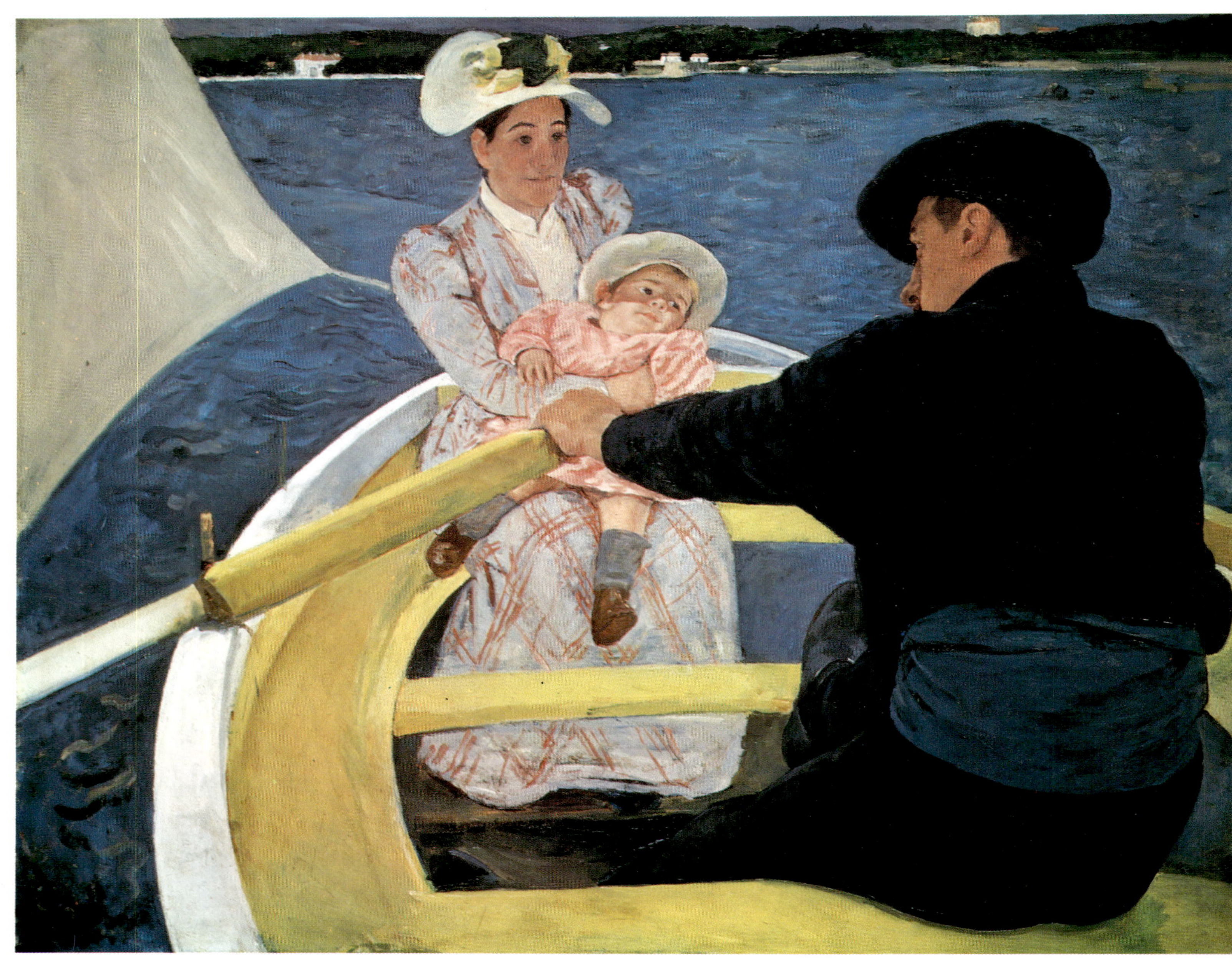

67. MARY CASSATT. *The Boating Party*. 1893–94. Canvas, 35 1/2 × 46 1/4″. Chester Dale Collection

Mary Cassatt

Mary Cassatt, the daughter of a wealthy banker, was born in what is now Pittsburgh, Pennsylvania, in 1844. She lived in Europe as a girl, and returned as a young woman to continue her art studies. Settling in Paris, she met Edgar Degas and through him joined the Impressionist group of painters. It was many years, however, until her reputation became secure with the public. Thanks to Cassatt's own purchases and her encouragement of American collectors, many fine Impressionist paintings found their way to this country. She died in France in 1926.

MARY CASSATT

The Boating Party

Child in a Straw Hat

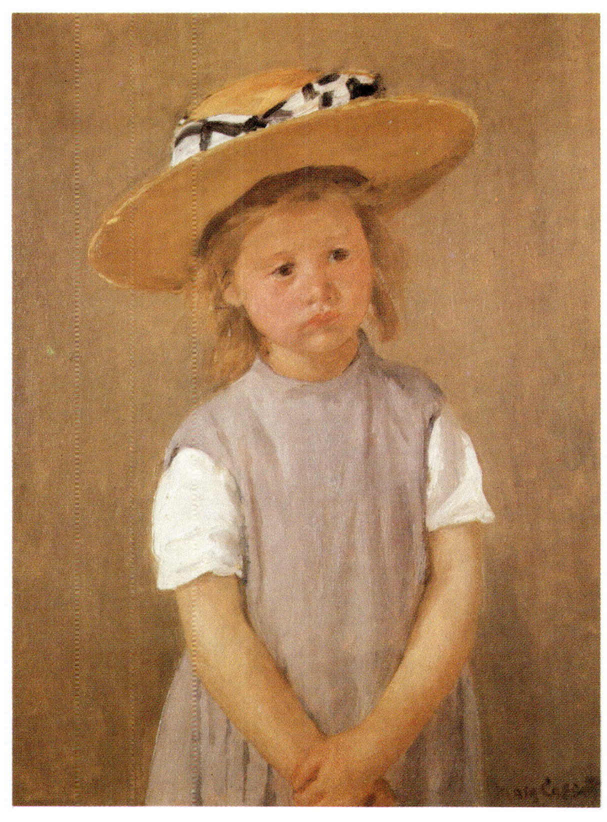

68. MARY CASSATT. *Child in a Straw Hat.* c. 1886. Canvas, 25 1/2 × 19 1/4″. Collection of Mr. and Mrs. Paul Mellon

The American artist Mary Cassatt's understanding of children is reflected in the two paintings shown here. The moods of the children are clearly conveyed: In *The Boating Party*, the indifferent child squirms in his mother's lap, while the little girl in the pinafore and large straw hat seems resigned and sad.

The Boating Party was painted on the French Riviera during the winter of 1893–94. Twenty years later Miss Cassatt wrote Durand-Ruel, a dealer who wished to buy the picture, "I do not wish to sell it. I have already promised it to my family. . . . I have so few things to leave my nieces and nephews!"

Apparently unaccustomed to boats, the mother sits stiffly, protectively clutching her restless child, who is determined to get off her lap. Notice that the artist has arranged her composition so that the oar, the man's arm, and the sheet of the sail all lead to the child. The forms are composed of almost flat shapes of varying colors; these shapes are locked together on the surface, further emphasizing the flatness, reducing the sense of space, and reminding us of the paintings of Manet (see plates 50, 51).

The artist has skillfully blended blue, blue-green, black, and, surprisingly, even a little pink to give movement to the water. Some of the blues in the boatman's sash echo the blues in the sea. The various greens in the woman's hat are found again in the sail, in the boat, and in the dark green of the shoreline. The pink touches here and there in the water pick up the pink in the baby's dress and in the plaid of the woman's costume.

Mary Cassatt painted the three-quarter-length picture *Child in a Straw Hat* about 1886, when living in the French countryside. This whimsical and appealing little French country child, in a high-necked pinafore—a mixture of grays, white, muted pinks, and lilac—over a white short-sleeved blouse, is shown against a grayish background. Her oversized yellow straw hat sits at a jaunty angle on her straggly blond hair.

The child's dark brown eyes look out of the picture, probably at the artist painting her. She is not an eager model, for her lips are pursed in a pout. Her casually crossed hands suggest her patience as she poses for her portrait. She would rather, no doubt, be playing in the summer sun with other children.

69. CHILDE HASSAM. *Allies Day, May 1917*. 1917. Canvas, 36 3/4 × 30 1/4″. Gift of Ethelyn McKinney in memory of her brother, Glen Ford McKinney

CHILDE HASSAM

Allies Day, May 1917

When America became active in World War I, Childe Hassam painted a series of some twenty-five scenes of the streets of New York decorated with the flags of the United States and its allies England and France. *Allies Day, May 1917*, considered by some the best of the series, appears to have been painted from an upper balcony of a building at the corner of Fifth Avenue and 52nd Street. Hassam was one of the first American painters to adopt the techniques of Impressionism, and *Allies Day* shows that he understood these methods well.

Though sketchily drawn, the buildings on the left side of the picture are clearly recognizable as St. Thomas's Episcopal Church, the University Club, and the Gotham Hotel. Buildings with balconies line the right side of the street. The glowing morning light illuminates the Gothic structure of the church. In the street, small black shapes suggest masses of people gathering for the parade. Dominating the scene are the gay fluttering flags of the Allies.

Hassam has made this painting so alive in its light, color, and movement that even though nothing is drawn clearly, we can easily imagine the marching soldiers, sailors, and marines, the stirring music, and the patriotic cheering of the crowds.

Childe Hassam

Childe Hassam was born in Dorchester, Massachusetts, in 1859. He was already doing illustrations for magazines and books when he began to paint. Hassam was especially interested in the temporary effects of light on familiar street scenes, and after his return to the United States often painted impressionistic views of New York. Hassam died in 1935, leaving a large collection of his pictures to be sold so that American and Canadian museums could purchase the work of other contemporary artists.

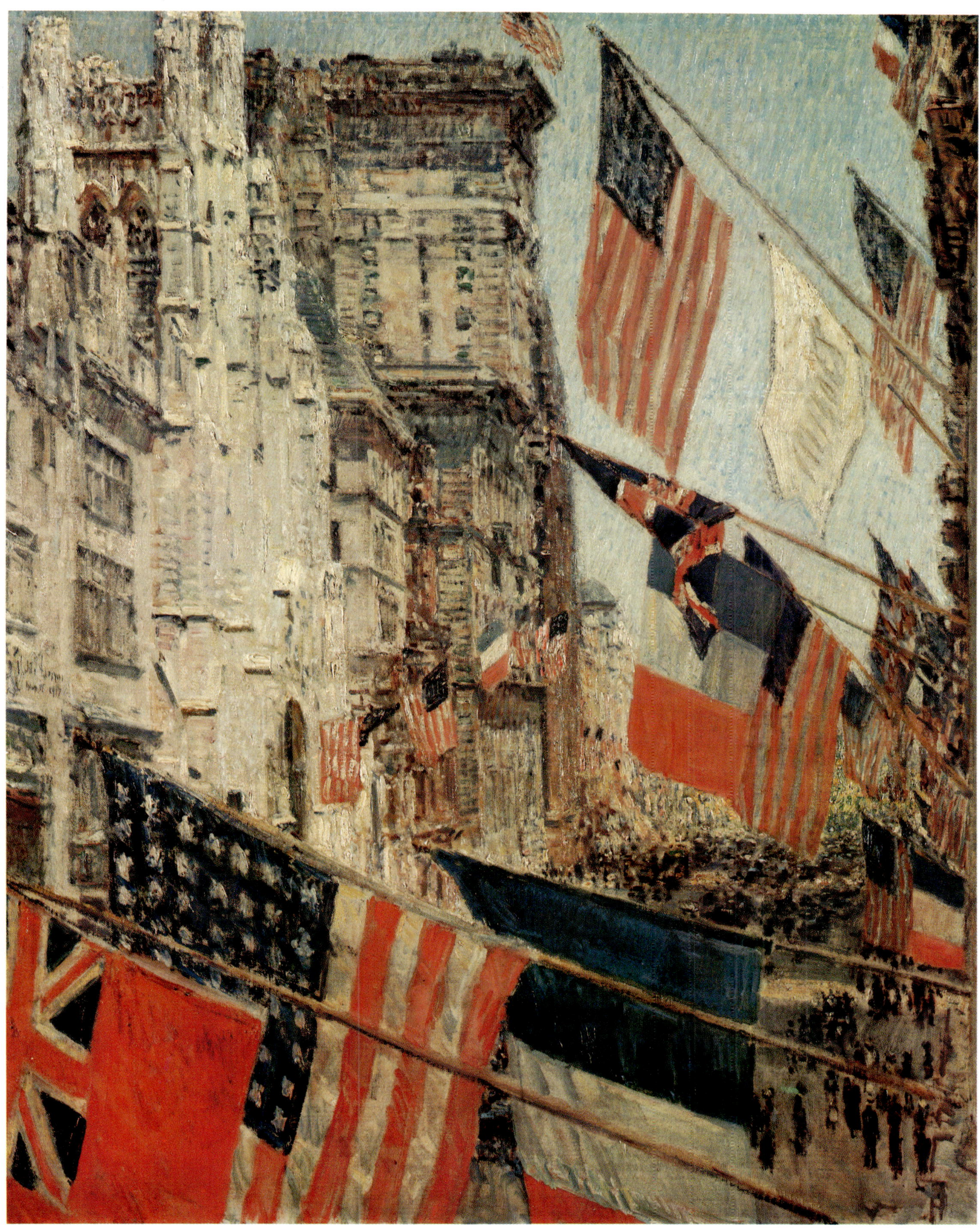

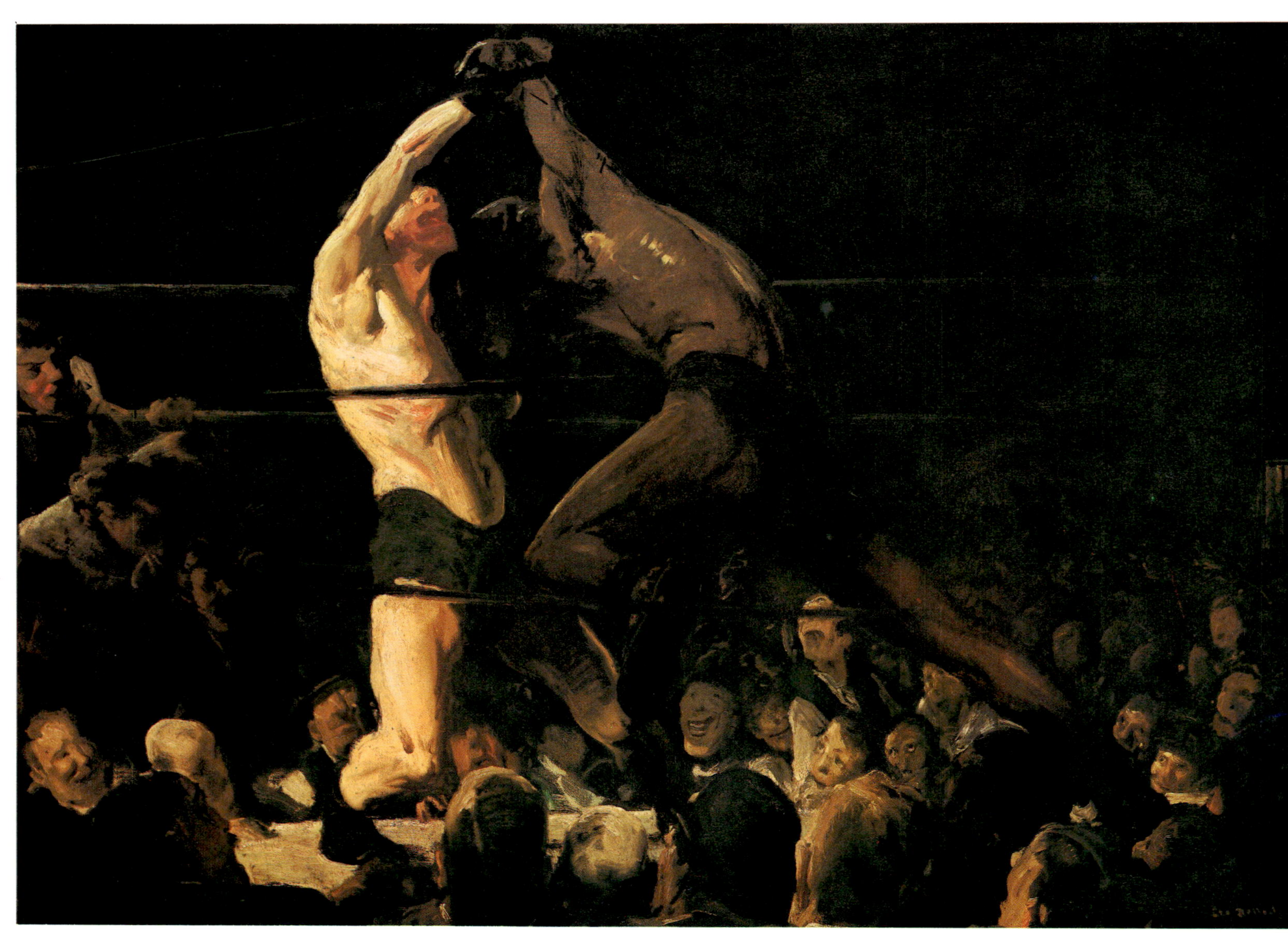

70. GEORGE BELLOWS. *Both Members of This Club*. 1909. Canvas, 45 1/4 × 63 1/8″. Chester Dale Collection

GEORGE BELLOWS

Both Members of This Club

George Bellows's great interest in boxing led him to attend the bout shown in this painting. The fight took place in 1909 at Tom Sharkey's Athletic Club near Bellow's studio in New York City. When Bellows painted *Both Members of This Club*, there was a New York State law forbidding public boxing. To get around the law, professional fighters were made members of private athletic clubs, where there was no restriction against bouts between members. Sometimes the boxers became members only for the night of the fight.

Spotlighted against the dark background are two struggling fighters, painted with unrelenting realism. Their bodies glisten with sweat. The white flesh of the boxer on the left is accented by the blood on his face. The strain and pull of his muscles are defined in his raised right arm, the bend of his knees, and the turn of his neck, as he seems about to crumple in defeat. Lunging at him from the right, partly merging with the background yet powerfully and skillfully brought to the fore, is the winning fighter, who moves in for the knockdown. Through the position of the black man's legs, the curve of his back, and the thrust of his raised arm, Bellows suggests the force and brutality of the sport. At the extreme left of the ring are the dim, hovering figures of the referee and the trainers.

Around the ring Bellows has painted the spectators. His portrayal of them reveals his attitude toward them, for he has emphasized their leering, taunting, ghastly countenances. They seem hypnotized and fascinated by the violence. The artist has made us feel we are in the smoke-filled, stuffy arena at this climactic moment of the match.

George Bellows

Born in Columbus, Ohio, in 1882, George Bellows achieved success as a painter only a few years after arriving in New York to study art. Sporting events and urban scenes of ordinary people going about their affairs particularly attracted Bellows as artistic subject matter. His painting *Stag Night at Sharkey's* is one of the most famous in American art. Bellows was adept at drawing and lithography in addition to painting, and made nearly two hundred prints. Bellows died at the height of his career in 1925, at the age of forty-two.

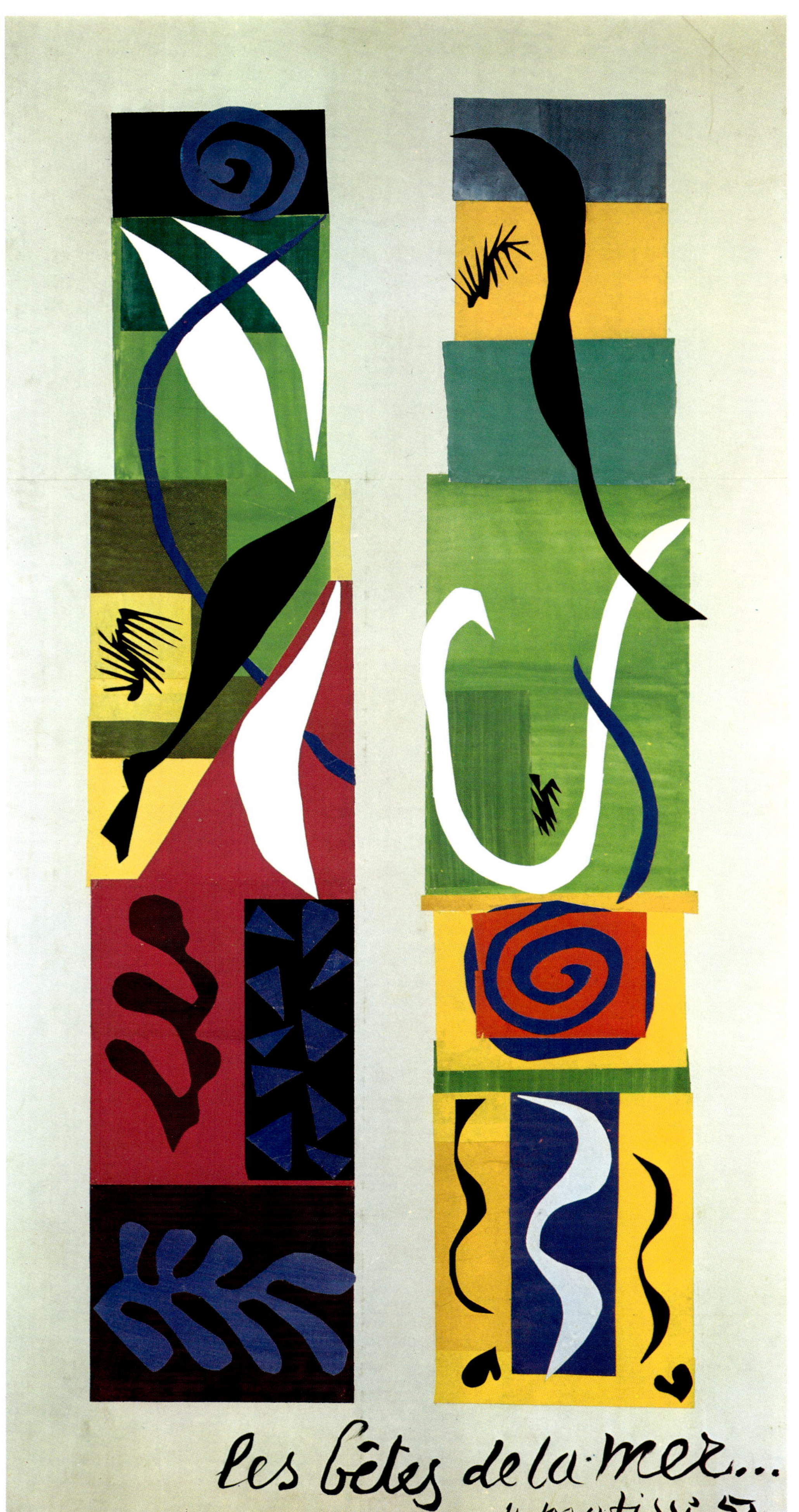
les bêtes de la mer...
H. Matisse 50

HENRI MATISSE

Les Bêtes de la Mer (Beasts of the Sea)

Woman with Amphora and Pomegranates

Henri Matisse created *Les Bêtes de la Mer*, or *Beasts of the Sea*, in 1950, toward the end of his long and successful career. The work, with its brilliant tropical colors and marine shapes, was perhaps inspired by the artist's 1930 trip to Tahiti.

The images in the picture reflect the organization of tropical sea life. Animals which feed from the ocean floor are represented in abstract and stylized form at the base of each column—coral and shellfish forms on the left, eels and snails on the right. Higher are swimming fish, including the spiky shapes of predatory tropical fish. The long, black, curving shape on the right may represent a sea horse. While the forms have been simplified, the sense of underwater life still remains. The work derives its spirit and strength from the arrangement of bold colors and shapes into decorative patterns and from the lyrical quality of line which always distinguishes Matisse's art.

The work is known as a *découpage*, the French term for paper cutouts. The forms were cut out of paper which had been painted by the artist, and then were pasted on a larger piece of canvas. The marks of the scissors can be seen on most of the forms, particularly on the white fish in the left column and the orange-and-blue snail and curving black form rising to the top in the right column.

In *Woman with Amphora and Pomegranates*, one of his late works, Matisse has used simple, flat forms to create a striking design. Although the form in the picture is recognizable as a woman, its elements are abstract units put together according to rules of balance and composition. The pomegranates are also very basic shapes; were they not named in the title, they could be taken to represent any fruit.

Matisse, then, was more concerned with the way the shapes could be arranged to make an appealing pattern than with what they were supposed to represent. Here, the shapes were placed on a small square which had been placed on a larger field of lighter paper. All the shapes overlap from the square onto the lighter area, but they are organized around the square, whose center provides the focus of the work. With the strong vertical emphasis provided by the band on the right, the viewer's eye is led outward and to the left, passing over and around the amphora, down to the pomegranates, and finally to the bottom of the band.

As in his other découpages, Matisse retained the marks his scissors made as he cut out the shapes. This emphasis on technique, making clear through the image the manner in which it was created, became an important tenet of twentieth-century art.

◁ 71. HENRI MATISSE. *Les Bêtes de la Mer (Beasts of the Sea)*. 1950. Paper on canvas (collage), 116 3/8 × 60 5/8″. Ailsa Mellon Bruce Fund

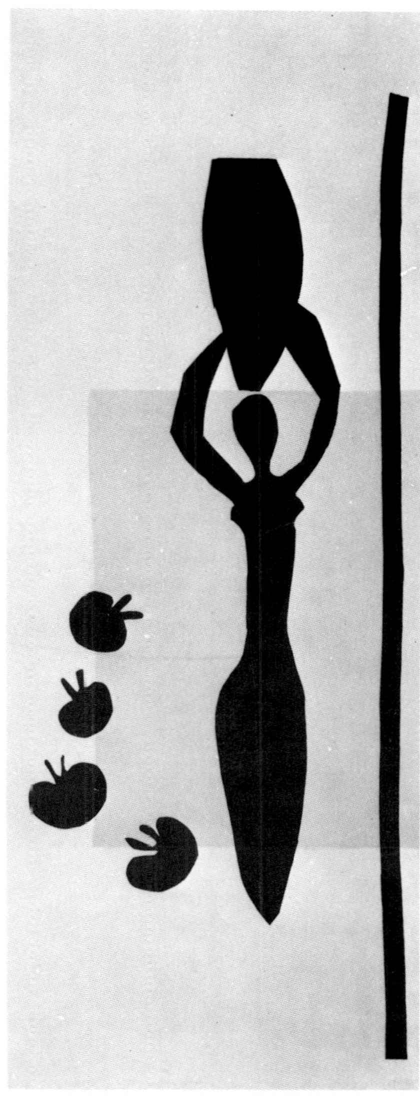

72. HENRI MATISSE. *Woman with Amphora and Pomegranates*. 1952. Paper on canvas (collage), 96 × 37 7/8″. Ailsa Mellon Bruce Fund

Henri Matisse

The French painter and sculptor Henri Matisse was born in 1869 at Le Cateau. He was trained as a lawyer, but while recuperating from an illness took up painting and soon abandoned all other pursuits forever. When his pictures began to reflect his interest in color for purely expressive purposes, Matisse rapidly gained a following of enthusiastic young artists. This group was called *fauves*, or "wild beasts," because their color was often garish and bore little relation to the way things actually look. In a few years, Matisse's work was shown all over Europe and in America. Among the artistic greats of the twentieth century, only Picasso is of equal stature. Matisse spent the last years of his life in southern France working chiefly on découpages. He died in 1954 at the age of eighty-five.

PABLO PICASSO

Family of Saltimbanques

In 1905 Pablo Picasso painted a series of circus pictures. The *Saltimbanques* is the most ambitious work of this group. It follows what was later called the artist's Blue Period, when he worked in various shades of blue, and in its somber mood recalls the paintings of this period.

Beneath a fog-streaked sky with an area of bright blue breaking through is a barren, indefinite landscape painted in subdued shades of pinks and browns. In this bleak area there is not a blade of grass, a flower, shrub, or tree.

Grouped together on the left are five figures. Like the people in Manet's *The Old Musician* (plate 50), they are involved in their thoughts and unaware of each other. They seem suspended in time—forever motionless. In the costume of Harlequin, whose profile is that of Picasso, the artist has used many shades of blue along with rose and black, tones he has repeated in touches of color on the face of the director of the troupe, a heavyset man carrying a bag, and in the skin and clothing of the little girl and two boys. The muted colors strike a chord of sadness and make us aware of the undercurrent of unhappiness which is always part of a clown's life.

On the right, separated from the group, a young woman is seated. Her blue blouse is partially covered by a gray-white shawl. Parts of her hat and reddish-pink skirt blend into the background. Near the woman, almost merging into the landscape, the outlines of a pitcher with a delicately drawn handle and fluted spout can be seen.

Picasso has used black accents and outlines to sharpen the effect of the subdued and subtle colors. He has heightened the feeling of isolation by placing the group on one side of the picture and the lone young woman on the other. Dressed in the costumes of their trade, which is supposed to bring amusement and pleasure, the figures seem alone, lost, forlorn.

Pablo Picasso

Born in Málaga, Spain, in 1881, Pablo Picasso was a child prodigy. His father, an art teacher, encouraged the gifted boy, and at sixteen Pablo successfully showed his work in Barcelona and published drawings in the Catalan journal *Joventut* ("Youth"). Drawn to the exciting art world of Paris, Picasso settled there and soon attracted a following among other young artists. Not long after, he invented Cubism, the most radical and innovative movement of this century, and went on to cross currents and styles in a way that left an indelible mark on the work of nearly every major artist of his time. Picasso, who was equally comfortable with painting, sculpture, collage, prints, stage and costume design, and ceramics, was acknowledged as a master and received innumerable honors before his death in 1973.

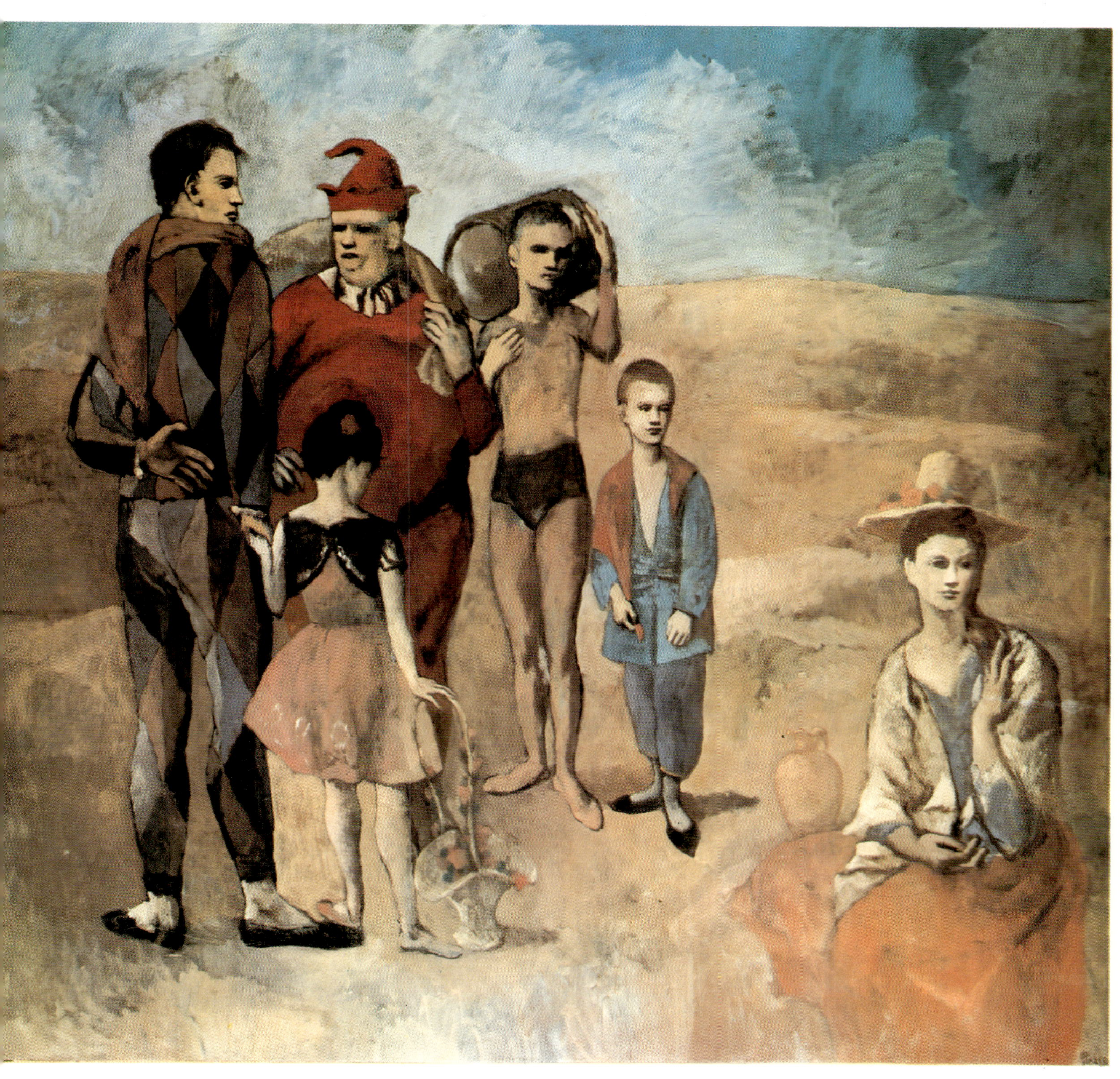

73. PABLO PICASSO. *Family of Saltimbanques*. 1905. Canvas, 83 3/4 × 90 3/8″. Chester Dale Collection

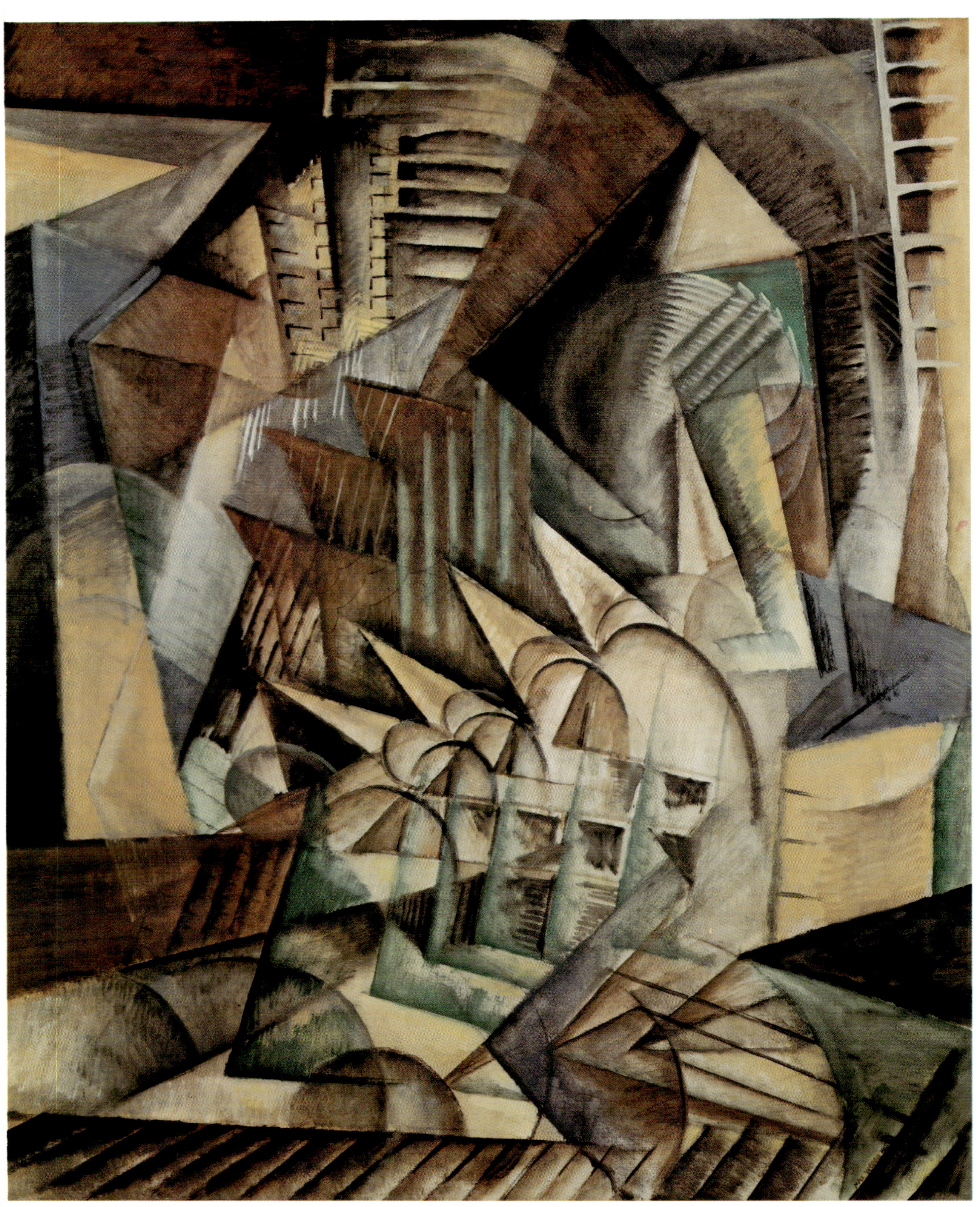

74. MAX WEBER. *Rush Hour, New York*. 1915. Canvas, 36 1/4 × 30 1/4″. Gift of the Avalon Foundation

MAX WEBER

Rush Hour, New York

In *Rush Hour, New York,* Max Weber, one of the foremost American painters of the early decades of the twentieth century, has used abstract, geometrical forms to express the movement, noise, and vibrancy of the great metropolis. The picture blends elements of two European styles: Cubism, breaking up objects into facets to show them from a number of different angles of vision at the same time, and Futurism, a movement which was concerned with portraying speed and objects in motion. The use of spiky forms which are linked by the forceful lines throughout the composition conveys the energy and vitality of the city. Weber expresses the city's diversity by juxtaposing rounded shapes and forms with angular ones. The forms are not entirely unrecognizable, but recall specific elements of the urban landscape: skyscrapers, subway trains, streets, flashing lights, hurrying people.

The colors in the painting are limited, reflecting the somber palette of early Cubism, when colors were restricted.

Max Weber

Born in Bialystok, Poland, in 1881, and raised in the tradition of Russian Judaism, Max Weber was ten when his family moved to New York. He studied art as a teenager and later taught before going to Europe to familiarize himself with contemporary artistic developments. Upon his return to the United States, Weber worked in the new styles he had discovered in Paris, and soon his Cubist-inspired pictures showed him to be a pioneer of American abstract painting. Weber died at the age of eighty in 1961.

JACKSON POLLOCK

Lavender Mist

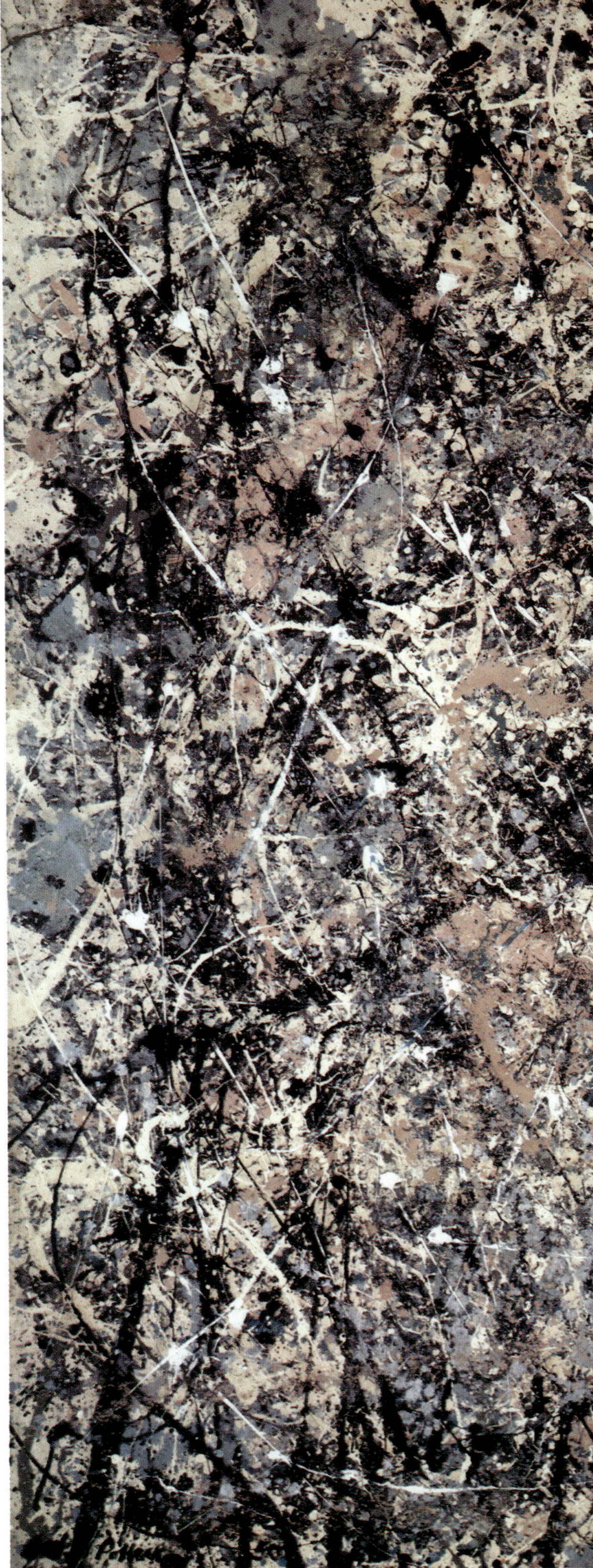

Lavender Mist is a large poured painting by Jackson Pollock, one of the major American Abstract Expressionist painters. The work was done in 1950, when the artist was at the height of his powers. It is totally abstract; that is, it contains no forms we can recognize. Its subtleties of pattern, color, and texture, however, make it evocative and powerful.

By the time he painted *Lavender Mist*, Pollock had perfected the technique of pouring paint from a can, often using brushes or sticks, onto a sheet of canvas laid on the floor. This is a totally different technique from that customarily used, where the canvas is placed on an easel or propped against a wall. In order to make a picture, Pollock would move around and around the piece of canvas, leaning far over it, sometimes even stepping onto it, to reach an area and apply the paint in long streams. The result, visible in *Lavender Mist*, shows an apparently random but actually carefully planned pattern of color and line which reflects the motions of the painter in pouring the paint, and gives the viewer a feeling of continuous movement and activity. Pollock said that for him, moving around and over the canvas as he painted made him feel as if he were actually in the painting as he created it. Some critics have called his style of painting, and that of other painters whose gestures, or motions, are reflected in their work, "Action Painting."

Lavender Mist combines exuberance and classical elegance. It presents a shimmering surface of delicate colors. By building up layer upon layer of entwining lines in varying colors, Pollock forces the viewer's eye to move along the lines over the surface of the painting and to lose itself in the richness of color and pattern created by the myriad lines. The composition is well balanced, with no area of the painting dominating another. This evenness of surface and design reveals Pollock's mastery of the pouring technique and shows the great care that went into creating the work.

Jackson Pollock

Born in 1912 in Cody, Wyoming, Jackson Pollock spent most of his youth in California. He never completed high school, but instead assisted in land surveys, traveled through the Southwest, and went to New York at seventeen to study art. During the Depression, he worked on the government's Federal Arts Project for artists in need of a job, painting dramatic, mystical pictures which showed the influence of American Indian art, Mexican mural painters, and contemporary artists like Picasso and Miró. Pollock's work grew increasingly turbulent and abstract until he finally began to drip and splatter paint directly onto canvases. Before his death in 1956, he had begun to stain shapes and figures onto canvas with black paint. His innovative techniques opened a new world of possibilities for American painters.

75. JACKSON POLLOCK. *Lavender Mist*. 1950. Canvas, 7′4″ × 9′11″. Ailsa Mellon Bruce Fund, 1976

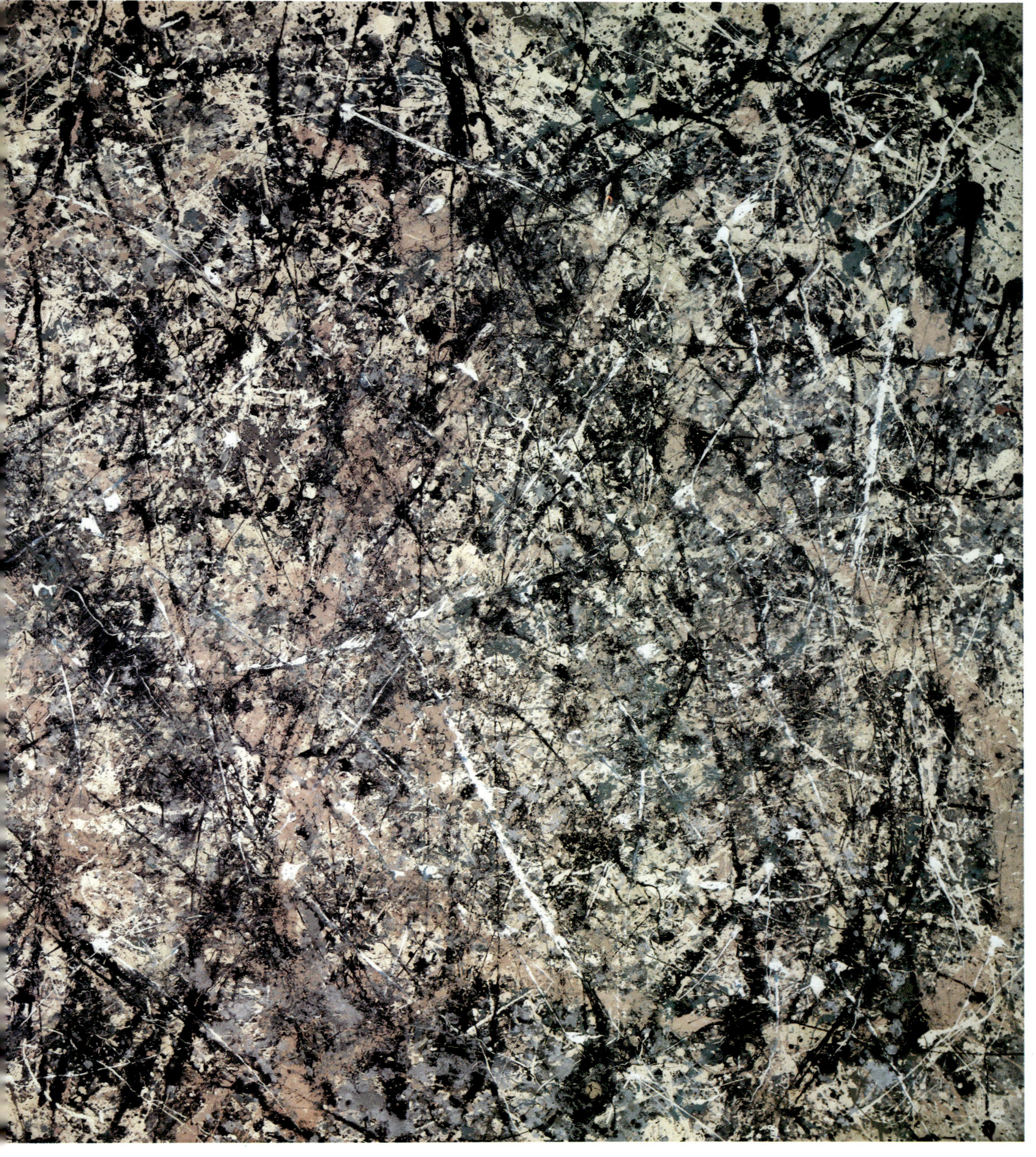

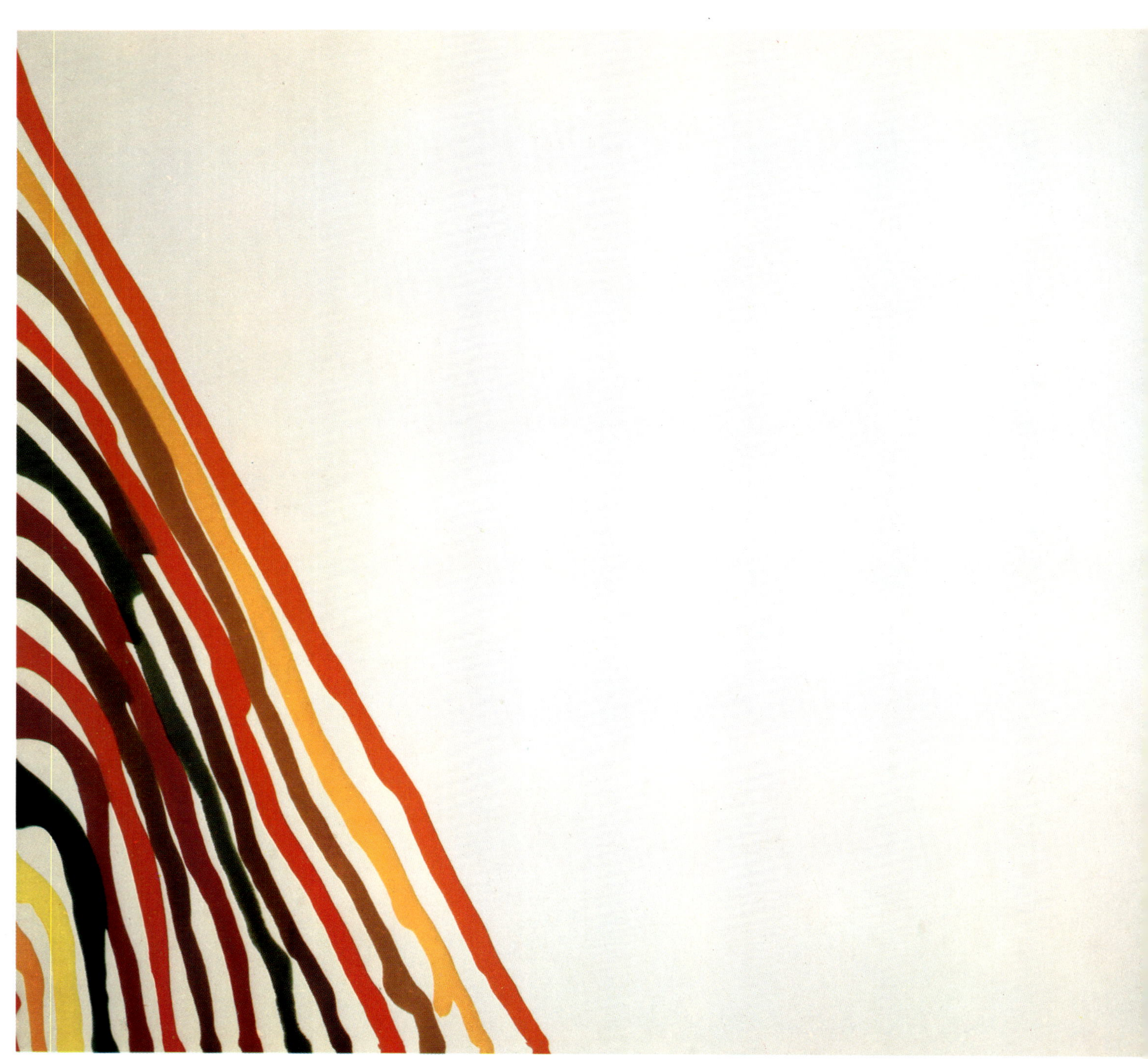

76. MORRIS LOUIS. *Beta Kappa*. 1961. Canvas, 103 1/4 × 173″. Gift of Marcella Louis Brenner

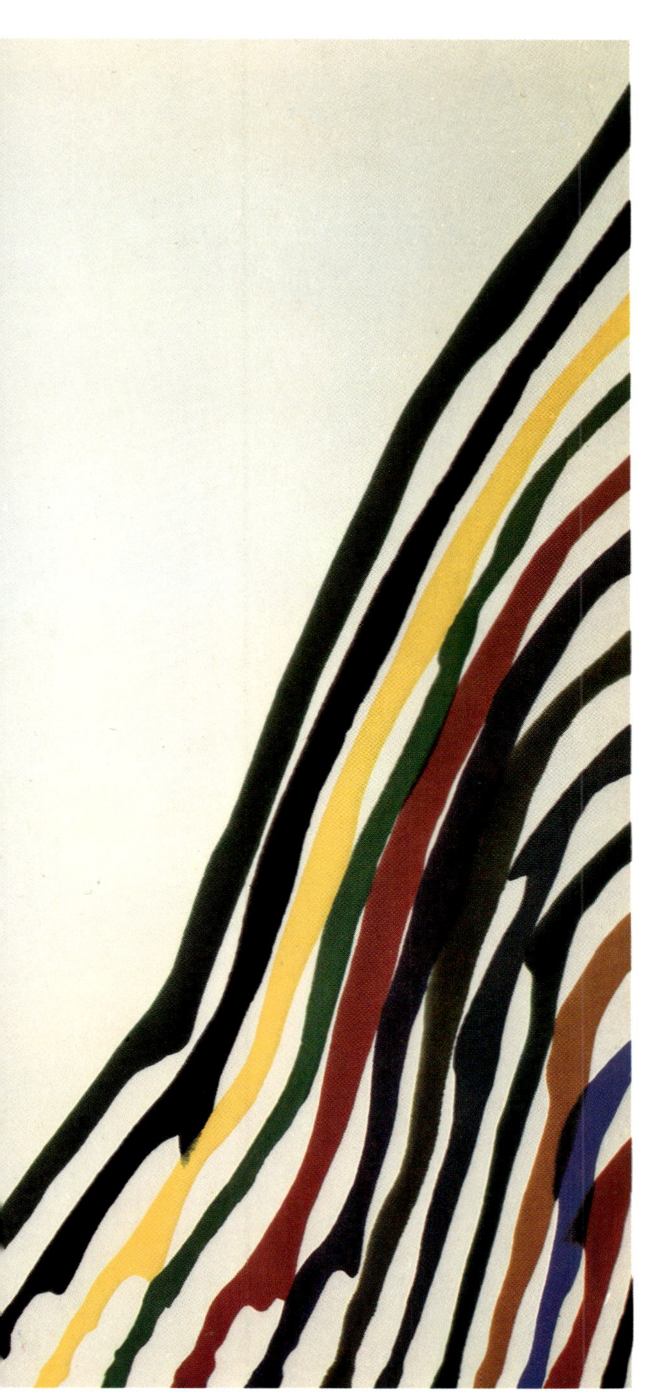

MORRIS LOUIS

Beta Kappa

Morris Louis's mature art, which, like Pollock's work, is abstract, is chiefly concerned with color. The painting *Beta Kappa* shows how Louis used color to make a rich and varied image with very simple elements.

The painting is made by a process called staining. To make a stain painting, pigment, or color, is soaked into the canvas, so that it actually becomes part of it. In *Beta Kappa* we see the falling rivulets of color and at the same time the tinted canvas, over and through which the color passes.

The painting derives its excitement from the balance of colors on each side of the canvas and from the tension created between the banks of color at the sides and the unpainted white field in the center. This configuration in Louis's work is called an "Unfurled." The idea of opening, or unfurling, comes from the sense of motion created by the color rivulets and from the action of the viewer's eye as it moves back and forth across the canvas. Traditionally painters have used diagonals at the sides of their works to imply distance and, often, motion. The diagonal color banks in *Beta Kappa* are used in the same way. By simplifying traditional elements of painting, Louis here has created a balance of color and shape which is appealing to the eye.

Beta Kappa is a very large painting. American painters of the 1950s and 1960s expanded the picture field to an immense size, so that the viewer's experience of the painting would be overwhelming.

Morris Louis

Morris Louis was born Morris Bernstein in 1912 in Baltimore. He studied painting in Maryland and later worked with other artists on the Federal Arts Project in New York. Louis returned to Baltimore to teach before moving to Washington, D.C., where he met the painter Kenneth Noland. Together, Louis and Noland began to search for a new way to make color the subject of painting, and they found their solution in a stain painting by Helen Frankenthaler they saw during a visit to New York. Louis began to pour streams of paint on his canvases to create beautiful shapes, stripes, or veils made up of pure color. These paintings won a position of prominence for Louis in the art world. He died in 1962.

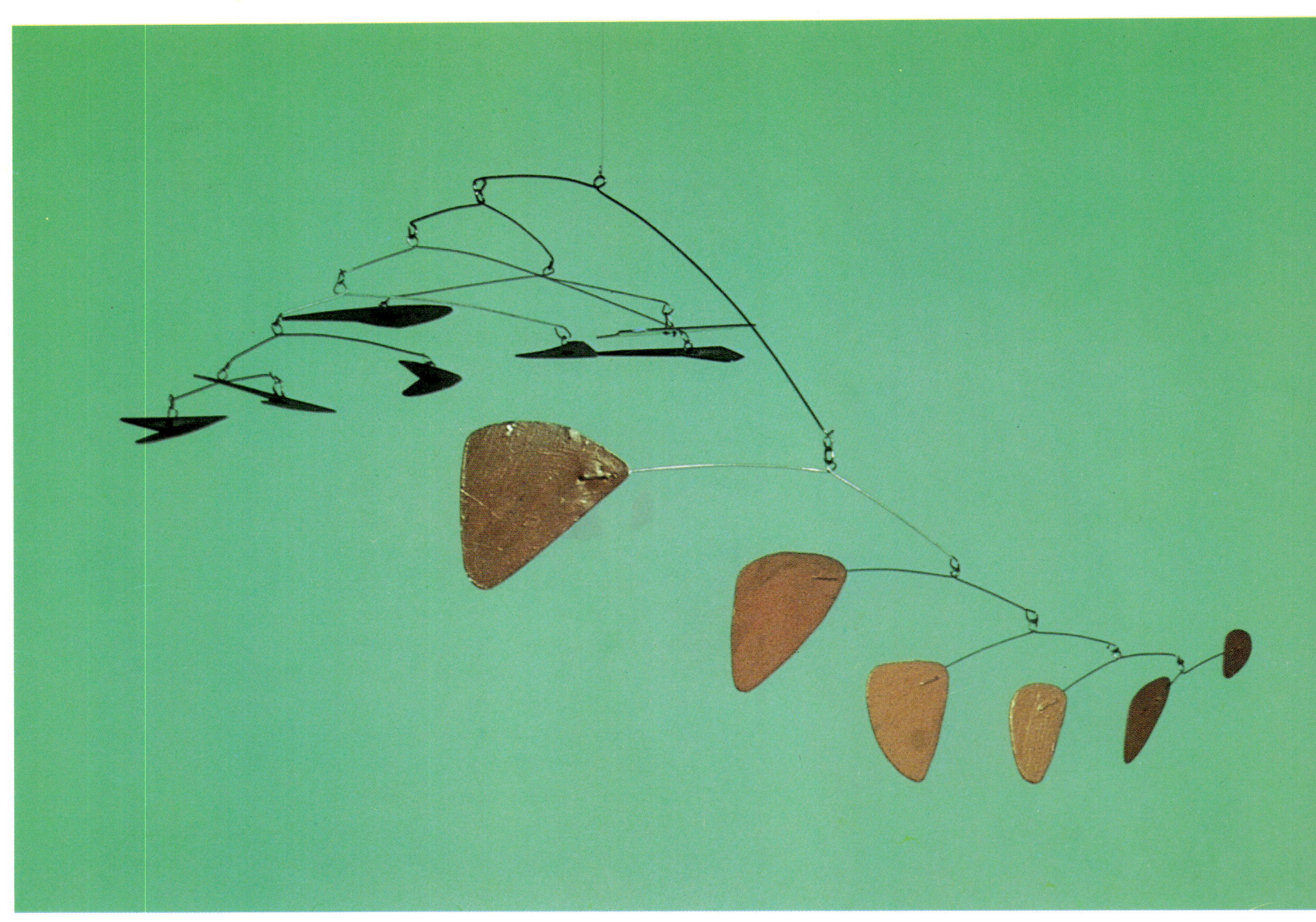

77. ALEXANDER CALDER. *Untitled mobile* (model shown above). 1972–76. Painted aluminum and steel, 30 × 76′. Collector's Committee Fund

ALEXANDER CALDER

Untitled Mobile

The mobile created by the late Alexander Calder for the multistoried central court of the National Gallery's East Building is an enlarged version of a kind of design he perfected in the 1930s.

In his mobiles, Calder used industrial materials, such as steel, basing his work on engineering and aesthetic principles, which he understood very well. The movements of Calder's mobiles are not programmed or timed, but depend on the chance movements of air currents. This combination of precise, certain structure and dependence on random air currents is one of the delights of the mobile. We never know precisely when, how much, or in what direction the forms will move.

The giant mobile designed for the East Building relates in its enormous scale to the great size of the interior court space. A smaller work would be lost in this space, but the size of this mobile actually makes the building seem less vast and more human in scale. This harmonizing of the sculpture with its setting is characteristic of Calder's work—of his stabiles as well as his mobiles.

The 930-pound mobile, one of Calder's largest works, is constructed of hollowed metal, like an airplane wing, which allows it to float lightly in the air. Its red, white, and blue shapes suggest wings and fins, forms which appear throughout the artist's work. The colors' strength and cheerfulness enliven the large space. As they swing about in the currents of air, the forms enchant us with their ever-changing rhythms.

Alexander Calder

Alexander Calder was born in Philadelphia in 1898. Though both his parents were artists, he intended to become an engineer, and only began painting seriously when he took night classes in New York City. On a trip to Paris he made his first wire sculptures of circus figures and animals, and soon incorporated movement into his works through the use of motors or handcranks. It was a short step to the creation of the mobiles for which he later became famous. Though Calder called Connecticut his home, he spent time each year in New York and France, where he not only painted and sculpted, but designed stage sets, prints, and tapestries. The artist died in 1976.

SELECTED BIBLIOGRAPHY

GENERAL

BIHALJI-MERIN, OTO. *Masters of Naive Art*. Trans. by Russell M. Stockman. New York: McGraw-Hill, 1971.

CAMPBELL, ANN. *Paintings: How to Look at Great Art*. New York and London: Franklin Watts, 1970.

HIND, ARTHUR M. *An Introduction to a History of Woodcut*. 2 vols. New York: Dover, 1964.

———. *History of Engraving and Etching*. 3rd rev. ed. New York: Dover, 1964.

JANSON, H. W. *History of Art*. 2nd ed. New York: Abrams, 1977.

——— and DORA JANE JANSON. *The Story of Painting for Young People*. New York: Abrams, 1962.

RAYNAL, MAURICE. *History of Modern Painting*. Geneva: Skira, 1956.

ROBB, DAVID M. *History of Painting*. New York: Harper, 1951.

TUFTS, ELEANOR. *Our Hidden Heritage: Five Centuries of Women Artists*. New York and London: Paddington Press, 1974.

EUROPEAN ART

BENESCH, OTTO. *The Art of the Renaissance in Northern Europe*. Rev. ed. London: Phaidon, 1965.

BLUNT, ANTHONY. *Art and Architecture in France 1500–1700*. Baltimore: Penguin, 1954.

BOIME, ALBERT. *The Academy and French Painting in the Nineteenth Century*. New York: Phaidon, 1970.

DE WALD, ERNEST T. *Italian Painting 1200–1600*. New York: Holt, Rinehart & Winston, 1961.

EVANS, JOAN. *English Art, 1307–1461*. Oxford: Clarendon Press, 1949.

GERSON, H. and E. H. TER KUILE. *Art and Architecture in Belgium, 1600–1800*. Baltimore: Penguin, 1960.

GILBERT, CREIGHTON. *History of Renaissance Art Throughout Europe: Painting, Sculpture, Architecture*. New York: Abrams, 1973.

HARTT, FREDERICK. *History of Italian Renaissance Art*. New York: Abrams, 1969.

HELD, JULIUS and DONALD POSNER. *Seventeenth and Eighteenth Century Art: Baroque & Rococo*. New York: Abrams, 1971.

HOFMANN, WERNER. *The Earthly Paradise: Art in the Nineteenth Century*. Trans. by Brian Battershaw. New York: Braziller, 1961.

LASSAIGNE, JACQUES. *Spanish Painting from Velazquez to Picasso*. Geneva: Skira, 1952.

LEVEY, MICHAEL. *Painting in Eighteenth Century Venice*. London: Phaidon, 1959.

LEYMARIE, JEAN. *Dutch Painting*. Geneva: Skira, 1956.

———. *Impressionism*. Geneva: Skira, 1955.

POPE-HENNESSY, JOHN. *Italian Renaissance Sculpture*. London: Phaidon, 1970.

REWALD, JOHN. *The History of Impressionism*. 4th rev. ed. New York: Museum of Modern Art, 1973.

ROSENBERG, JAKOB, SEYMOUR SLIVE and E. H. TER KUILE. *Dutch Art and Architecture 1600–1800*. Baltimore: Penguin, 1966.

SEYMOUR, CHARLES, JR. *Sculpture in Italy 1400 to 1500*. Baltimore: Penguin, 1966.

STECHOW, WOLFGANG. *Dutch Landscape Painting of the Seventeenth Century*. Rev. ed. London: Phaidon, 1966.

STONE, LAWRENCE. *Sculpture in Britain: The Middle Ages*. 2nd ed. Baltimore: Penguin, 1972.

WATERHOUSE, ELLIS K. *Italian Baroque Painting*. New York and Greenwich: New York Graphic Society and Phaidon, 1963.

———. *Painting in Britain*. Baltimore: Penguin, 1962.

WITTKOWER, RUDOLF. *Art and Architecture in Italy, 1600–1750*. 3rd ed. Baltimore: Penguin, 1973.

AMERICAN ART

BEARDEN, ROMARE and HARRY HENDERSON. *Six Black Masters of American Art*. New York: Zenith, 1972.

FLEXNER, JAMES THOMAS. *That Wilder Image*. Boston: Little, Brown, 1962.

LARKIN, OLIVER. *Art and Life in America*. Rev. ed. New York: Holt, Rinehart & Winston, 1960.

LIPMAN, JEAN. *American Primitive Painting*. New York: Oxford, 1945.

——— and ALICE WINCHESTER. *Primitive Painters in America, 1750–1950*. New York: Dodd, Mead, 1950.

———. *The Flowering of American Folk Art, 1776–1876*. New York: Viking and Whitney Museum of American Art, 1974.

POUSETTE-DART, NATHANIEL. *Leaders of American Impressionism*. Brooklyn, N. Y.: The Brooklyn Museum, 1937.

RICHARDSON, EDGAR P. *Painting in America*. New York: Crowell, 1965.

MONOGRAPHS

BELLOWS: BRAIDER, DONALD. *George Bellows and the Ashcan School of Painting*. New York: Doubleday, 1971.

YOUNG, MAHONRI SHARP. *The Paintings of George Bellows*. New York: Watson-Guptill, 1973.

BONNARD: Fermigier, André. *Pierre Bonnard.* New York: Abrams, 1969.

CALDER: Calder, Alexander. *Calder: An Autobiography with Pictures.* New York: Pantheon, 1966.

Sweeney, James J. *Alexander Calder.* New York: Solomon R. Guggenheim Museum, 1964.

CANALETTO: Constable, W. G. *Canaletto.* 2 vols. Oxford: Clarendon Press, 1962.

Feles, Adrian. *Canaletto.* London: Paul Hamlyn, 1967.

CASSATT: Breeskin, Adelyn D. *Mary Cassatt: A Catalogue Raisonné.* Washington, D.C.: Smithsonian Institution Press, 1970.

Carson, Julia. *Mary Cassatt.* New York: McKay, 1966.

CASTAGNO: Richter, George M. *Andrea del Castagno.* Chicago: University of Chicago Press, 1943.

CATLIN: Haverstock, Mary S. *Indian Gallery: The Story of George Catlin.* New York: Four Winds, 1973.

McCracken, Harold. *George Catlin.* New York: Dial, 1959.

CÉZANNE: Cézanne, Paul. *Paul Cézanne.* Ed. by Meyer Schapiro. 3rd ed. New York: Abrams, 1962.

Elgar, Frank. *Cézanne.* New York: Praeger, 1974.

CHARDIN: Rosenberg, Pierre. *Chardin.* Trans. by Helga Harrison. Geneva: Skira, 1963.

Wildenstein, Georges. *Chardin.* Rev. by Daniel Wildenstein, trans. by Stuart Gilbert. Oxford: Cassirer, 1969.

CONSTABLE: Fleming-Williams, Ian. *Constable's Landscapes, Watercolours and Drawings.* London: The Tate Gallery, 1976.

Reynolds, Graham. *Constable, the Natural Painter.* New York: McGraw-Hill, 1965.

COPLEY: Flexner, James Thomas. *John Singleton Copley.* Boston: Houghton Mifflin, 1948.

Prown, Jules David. *John Singleton Copley.* Cambridge: Harvard University Press, 1966.

COROT: Leymarie, Jean. *Corot.* Trans. by Stuart Gilbert. Geneva: Skira, 1966.

CRANACH: Ruhmer, Eberhard. *Cranach.* London: Phaidon, 1963.

DÜRER: Panofsky, Erwin. *Albrecht Dürer.* 3rd ed., 2 vols. Princeton: Princeton University Press, 1948.

FRA ANGELICO: Argan, Giulio Carlo. *Fra Angelico.* Trans. by James Emmons. Geneva: Skira, 1955.

Pope-Hennessy, John. *Fra Angelico.* New York: Phaidon, 1952.

———. *Fra Angelico.* Rev. ed. Ithaca: Cornell University Press, 1974.

FRAGONARD: Thullier, Jacques. *Fragonard.* Trans. by Robert Allen. Geneva: Skira, 1967.

Wildenstein, Georges. *The Paintings of Fragonard.* Trans. by C. W. Chilton. New York: Phaidon, 1960.

GAUGUIN: Bowness, Alan. *Gauguin.* London and New York: Phaidon and Praeger, 1971.

Goldwater, Robert. *Paul Gauguin.* New York: Abrams, 1957.

Rewald, John. *Paul Gauguin.* New York: Abrams, 1952.

GÉRICAULT: Berger, Klaus. *Géricault and His Work.* Trans. by Winslow Ames. Lawrence: University of Kansas Press, 1955.

Eitner, Lorenz. *Géricault.* Los Angeles: Los Angeles County Museum of Art, 1971.

GOYA: Harris, Enriqueta. *Goya.* London: Phaidon, 1972.

GRECO, EL: Lassaigne, Jacques. *El Greco.* London: Thames and Hudson, 1973.

Wethey, Harold E. *El Greco and His School.* 2 vols. Princeton: Princeton University Press, 1962.

HASSAM: Pousette-Dart, Nathaniel. *Childe Hassam.* New York: Frederick A. Stokes, 1922.

HOLBEIN: Langdon, Helen. *Holbein.* Oxford and New York: Phaidon and E. P. Dutton, 1976.

HOMER: Gardner, Albert T. E. *Winslow Homer.* New York: Clarkson Potter, 1961.

Goodrich, Lloyd. *Winslow Homer.* New York: Braziller, 1959.

LEGROS: Wright, Harold J. L. *The Etchings, Drypoints and Lithographs of Legros.* London: Print Collectors' Club, 1934.

LEONARDO DA VINCI: Baroni, Constantino. *All the Paintings of Leonardo da Vinci.* Trans. by Paul Colacicchi. New York: Hawthorn, 1961.

Goldscheider, Ludwig. *Leonardo da Vinci: Life and Work.* 7th ed. London: Phaidon, 1964.

Pater, Walter. *Leonardo da Vinci.* London: Phaidon, 1972.

LOUIS: Carmean, E. A., Jr. *Morris Louis: Major Themes and Variations.* Washington, D.C.: National Gallery of Art, 1976.

Fried, Michael. *Morris Louis.* New York: Abrams, 1971.

MANET: Bataille, Georges. *Manet.* Trans. by A. Wainhouse and J. Emmons. Geneva: Skira, 1955.

Hanson, Anne Coffin. *Manet and the Modern Tradition.* New Haven: Yale University Press, 1976.

MATISSE: Barr, Alfred H., Jr. *Matisse, His Art and Public.* New York: Museum of Modern Art, 1951.

Jacobus, John. *Henri Matisse.* New York: Abrams, 1973.

Russell, John. *The World of Matisse, 1869–1954.* New York: Time-Life, 1969.

PICASSO: Barr, Alfred H., Jr. *Picasso: Fifty Years of His Art.* New York: Museum of Modern Art, 1946.

Duncan, David Douglas. *Picasso's Picassos.* New York: Harper, 1961.

Penrose, Roland. *Picasso.* 2nd ed. London and New York: Phaidon and Praeger, 1972.

PIERO DI COSIMO: Douglas, R. Langton. *Piero di Cosimo.* Chicago: University of Chicago Press, 1946.

POLLOCK: O'Hara, Frank. *Jackson Pollock.* New York: Braziller, 1959.

Robertson, Bryan. *Jackson Pollock.* New York: Abrams, 1960.

RAPHAEL: Beck, James A. *Raphael.* New York: Abrams, 1976.

Fischel, Oskar. *Raphael.* Trans. by Bernard Rackham. 2 vols. London: Kegan Paul, 1948.

REMBRANDT: Benesch, Otto. *Collected Writings: Essays on Rembrandt.* Vol. 1. Ed. by Eva Benesch. New York: Phaidon, 1970.

Rosenberg, Jakob. *Rembrandt, Life and Works.* London and New York: Phaidon and Praeger, 1964.

ROUSSEAU: Rich, Daniel Catton. *Henri Rousseau.* New York: Museum of Modern Art and Simon and Schuster, 1946.

RUBENS: Wedgewood, Cicely V. *The World of Rubens.* New York: Time-Life, 1967.

White, Christopher. *Rubens and His World.* New York: Viking, 1968.

STUART: Flexner, James Thomas. *Gilbert Stuart.* New York: Knopf, 1955.

Richardson, Edgar P. *Gilbert Stuart, Portraitist of the Young Republic.* Providence: Rhode Island School of Design, 1967.

TURNER: Rothenstein, John and Martin Butlin. *Turner.* New York: Braziller, 1964.

Lindsay, Jack. *J. M. W. Turner: A Critical Biography.* New York: New York Graphic Society, 1966.

VAN DYCK: Puyvelde, Leo Van. *Van Dyck.* Amsterdam: Elsevier, 1950.

VAN GOGH: Hammacher, A. M. *Genius & Disaster: The Ten Creative Years of Vincent van Gogh.* New York: Abrams, 1968.

Wallace, Robert. *The World of Van Gogh.* New York: Time-Life, 1969.

WATTEAU: Brookner, Anita. *Watteau.* London: Paul Hamlyn, 1967.

Schneider, Pierre. *The World of Watteau.* New York: Time-Life, 1967.

WEBER: Werner, Alfred. *Max Weber.* New York: Abrams, 1975.

INDEX

All numbers in this index refer to plate numbers; those in *italic* type denote colorplates.

ACKNOWLEDGMENTS

I wish to take this opportunity to express my sincere appreciation to the following people at the National Gallery of Art:

Paul Mellon, President, for his generous cooperation; J. Carter Brown, Director, for his gracious foreword; Charles Parkhurst, Assistant Director and Chief Curator, for his splendid guidance; Joseph G. English, Administrator, and Theodore S. Amussen, Editor-in-Chief, who were instrumental in paving the way; David Scott, Planning Consultant to the Director, and his assistant Carolyn B. Ganley, for their stimulating help; Margaret I. Bouton, Curator in Charge of Education, for her tireless assistance and support; Ruth R. Perlin, Curator/Education Specialist, Department of Extension Programs; William J. Williams, Staff Lecturer; and Jacqueline Sheehan, Graphic Arts, for sharing their knowledge with me.

I am also very grateful for the constructive assistance of David Alan Brown, Curator of Early Italian Painting at the Gallery; Douglas Lewis, Jr., Curator of Sculpture; David E. Rust, Curator of French Painting; Arthur Wheelock, Curator of Dutch and Flemish Painting; John Hand, Curator of Northern European Painting to 1700; E. A. Carmean, Jr., Curator of Twentieth Century Art; Elizabeth Foy, Administrative Assistant to the Director; the Chief Librarian, J. M. Edelstein, and his staff; Elise V. H. Ferber, Head, Art Information Service; Ira Bartfield, Coordinator of Photography; Max Leason, Manager of Publications; Stephen McEvitt, a member of his staff; Melanie Ness, Graphic Assistant to the Editor; Mary Dyer, Information Specialist; Jo Ann Purnell, Secretary to the Administrator; Mary Jane Pagan and Maria Mallus, secretaries in the Assistant Director's office, and Elizabeth Fletcher, secretary to David Scott, Planning Consultant to the Director.

This acknowledgment list would be incomplete if I did not mention the late William P. Campbell, Curator of American Painting at the National Gallery. Words are inadequate to express my profound appreciation to him; it has been a great honor to have been able to work with him over the years and especially on this volume.

A special expression of gratitude is due to the following outside of the Gallery for their valuable assistance: Adelyn D. Breeskin, Mollie B. Zion, Ann G. Yurow, Carolyn H. Wells, Martha M. Wright, and Katherine M. Ratzenberger.

Finally, my grateful thanks to Newton Pincus, Margaret Kaplan, and Patricia Gilchrest of Harry N. Abrams, publishers, who have been instrumental in making this volume possible.

WASHINGTON, D.C. MARIAN KING